Math
Hysteria

Math Hysteria

Fun and Games with Mathematics

IAN STEWART

OXFORD
UNIVERSITY PRESS

Great Clarendon Street, Oxford OX2 6DP

Oxford University Press is a department of the University of Oxford.
It furthers the University's objective of excellence in research, scholarship,
and education by publishing worldwide in

Oxford New York

Auckland Bangkok Buenos Aires Cape Town Chennai
Dar es Salaam Delhi Hong Kong Istanbul Karachi Kolkata
Kuala Lumpur Madrid Melbourne Mexico City Mumbai Nairobi
São Paulo Taipei Tokyo Toronto

Oxford is a registered trade mark of Oxford University Press
in the UK and in certain other countries

Published in the United States by
Oxford University Press Inc., New York

British Library Cataloguing in Publication Data

Data available

Library of Congress Cataloging in Publication Data

Data available

ISBN 978-0-19-861336-7

9

Typeset in Palatino by George Hammond Design
Printed in Great Britain by
Clays Ltd, St Ives plc

Contents

Preface vii

Figure Acknowledgements xi

1   I Know That You Know That... 1

2   Domino Theories. 11

3   Turning the Tables. 25

4   The Anthropomurphic Principle 37

5   Counting the Cattle of the Sun 47

6   The Great Drain Robbery. 57

7   Two-Way Jigsaw Puzzles 73

8   Tales of the Neglected Number 85

9   Is Monopoly Fair? 95

10 Monopoly Revisited 107

11 A Guide to Computer Dating 117

12 Dividing the Spoils 127

13 Squaring the Square 143

14 The Bellows Conjecture 155

15 Purposefully Piling Pyramids. 165

16   Be a Dots-and-Boxes Grandmaster 177

17   Choosily Chomping Chocolate. 187

18   Shedding a Little Darkness 199

19   Preposterous Piratical Predicaments 209

20   Million-Dollar Minesweeper. 217

Further Reading 227

Index. 231

Preface

When I was about 16 years old, one of the highlights of the month was reading Martin Gardner's 'Mathematical Games' column in *Scientific American*. Every column contained something new to attract my attention, and it was mathematical, and it was fun. I was fortunate in having some excellent mathematics teachers, so I already knew that mathematics was something you could enjoy, and that it was not cast in tablets of stone. Martin Gardner's column reinforced those messages. And even though the column was about games (later – I don't know why – it became 'Mathematical Recreations', which sounds stuffier), there was plenty of 'serious' mathematics mixed in with the fun.

It is probably fair to say that Martin Gardner's column was one of the reasons I ended up becoming a mathematician. It kept me interested and made it clear that there was plenty of room for new ideas and creative thinking. And unlike many of my fellow professionals, I never bothered to disentangle the 'serious' aspects of mathematics from the 'fun' ones. I don't mean that I couldn't see the distinction; I just mean that I didn't consider it to be terribly important. What mattered to me was the mathematics, and I enjoyed mathematical work and mathematical play without feeling any need to separate them.

In *The Colossal Book of Mathematics*, Martin Gardner says that his 'long and happy relationship with *Scientific American* ... began in 1952 when I sold the magazine an article on the history of logic machines.' After 25 years in the saddle, he decided to move on to other things, and his column was up for grabs. Douglas Hofstadter, Pulitzer prize-winning author of *Gödel, Escher, Bach, an Eternal Golden Braid*, was the first. He retitled the column 'Metamagical Themas', a clever anagram of 'Mathematical Games'. Then Kee Dewdney, author of *The Planiverse*, took up the reins, and the column became 'Computer Recreations'. At that point, the God of Mathematics Columns decided to open up an opportunity for me to get in on the act, although it took a while before this deity's intervention became apparent.

It was the French that started it. *Scientific American* is translated into more than a dozen languages; among them, French. 'Translated' isn't quite the word, because each foreign language edition includes its own material, and articles sometimes get moved from one month to another, or omitted altogether. The French edition is called *Pour La Science*, and its editor, Philippe Boulanger, wanted to keep the Mathematical Recreations column going in addition to running its Computer Recreations replacement. So he persuaded several French mathematicians to contribute mathematical columns. That worked for a few years, until the most regular contributor decided that he couldn't carry on any longer. A series of coincidences led to me being invited to take over, which I did, with alacrity. My first column appeared in September 1987. After a few years, the column spread to the German, Spanish, Italian, and Japanese editions of the magazine. In December 1990, a few months after Computer Recreations had metamorphosed back into Mathematical Recreations, I took over that slot in the parent magazine.

I also had a long and happy relationship with *Scientific American*, writing 96 columns over an 11-year period. I wrote a further 57 for *Pour La Science* and other translations; some in the four years before I began to contribute to the parent magazine, and some to turn what was initially a bimonthly column in America into a monthly one in France. Some of those columns have already been collected into books, a tradition also begun by Gardner: in English they are *Game, Set and Math* and *Another Fine Math You've Got Me Into*. ('Math' usually works better than 'Maths', and the magazine *is* called 'Scientific *American*'.) There are other collections in French and German. Eventually, I hope that every column will appear in at least – and preferably at most – one book. *Math Hysteria* is the next step in that programme, containing 20 columns hitherto unavailable in book form.

Martin Gardner is an impossible act to follow. There was never any prospect of his successors repeating the magic Gardner formula, and I'm pretty sure that none of us tried. I know I didn't. What we did try to do was replicate the spirit of the column: to present significant mathematical ideas in a playful mood. More than 3000 years ago the mathematics teachers of ancient Babylon put puzzles into their cuneiform texts to secure their pupils' attention. The ancient Egyptians did the same. I suspect that it was the Greeks, with their emphasis on high culture, that started the opposite tradition of presenting mathematics in a solemn, formal framework. I blame Euclid and his imitators for making mathematics tedious and mechanical, obsessed with checking that statement 17 of Theorem 46 follows from Lemma 25 and statement 18 follows from Proposition 12. I have nothing against proofs, but there is a time and a place, and the early development of visual intuition in mathematics is neither of those.

The chapters are arranged in no particular order, and you can dip in almost anywhere, although the two chapters on probability theory applied to Monopoly form a mini-series and are best read as such. The topics range from quirks of logic (I Know That You Know That ...) through combinatorics (Squaring the Square), curious numbers (Counting the Cattle of the Sun), and Geometry (Two-Way Jigsaw Puzzles), to more advanced topics including optimization (The Great Drain Robbery), and polyhedrons (The Bellows Conjecture). Some are about winning strategies in mathematical games (Choosily Chomping Chocolate) or the convoluted protocols of envy-free division (Dividing the Spoils) or impossibility proofs (Domino Theories). A few touch on practical matters: The Anthropomurphic Principle reveals why toast always falls buttered side down in a sensibly constructed universe, A Guide to Computer Dating explains why every culture has its own calendar and how they are related, and Purposefully Piling Pyramids figures out how many workers were needed to build the Great Pyramid of Khufu. And if you want to win a million dollars by thinking about (*not* by playing) computer games, there's Million-Dollar Minesweeper, linking the Windows operating system to the frontiers of mathematical research in the twenty-first century.

A word of thanks – no, words of gushing gratitude too profuse to record here – to cartoonist Spike Gerrell, whose mad cows, preposterous pirates, and perplexed monks enhance these pages. Spike has captured the spirit of the book with an insight and precision that I find astonishing. Thanks, too, to Oxford University Press and its staff of publishers, editors, copy-editors, and everything else that turns a vague idea into a finished book.

I must end by confessing that there's plenty of 'serious' mathematics sneaked in amid the fun and games – the most blatant examples have been hauled out into self-contained 'boxes' so that you won't feel cheated. So you can rest assured that while you're contemplating the strange antics of Archimedes' cows, you're also coming to grips with the fundamentals of number theory. However, I'm not trying to *teach* you anything. I just want you to enjoy a few samples from the remarkable human invention that is Mathematics.

IAN STEWART
Coventry, June 2003

Figure Acknowledgements

Figure 30: *American Mathematical Monthly*, reprinted with permission of the author and publisher.

Figures 22a, 23b, 23c, 25, 26, 27a, 27b, 27c, 29, 30, 31, and 32: *Dissections: Plane and Fancy*, 1997, Cambridge University Press, Greg N. Frederickson 1997, reprinted with permission of the author and publisher.

Figures 23a, 24, and 31b: *Journal of Recreational Mathematics*, reprinted with permission of the author and publisher.

Figure 27d: *Recreational Mathematics Magazine*, reprinted with permission of the author and publisher.

1

I Know
That You Know That...

Sometimes it's not enough merely to know something — you have to know that someone else knows. Or that they know that you know that they know that ... These considerations lead to the concept of 'common knowledge', and it makes a difference. Once something has become common knowledge, it becomes possible to make deductions about other people's reasoning.

The **extremely polite monks of the Perplexian order** like to play logical tricks on one another. One night, when Brothers Archibald and Benedict are asleep in their cell, Brother Jonah sneaks in and paints a blue blob on the top of each of their shaven heads. When they wake up, each of them of course notices the blob on the head of the other, but being polite, says nothing. Each vaguely wonders if he, too, has a blob, but is too polite to ask. Then Brother Zeno, who has never quite learned the art of tact, enters, and begins giggling. Upon being questioned, he remembers his manners, and refuses to say any more than 'At least one of you has a blue blob on his head.'

Of course, the monks both know that. But then Archibald starts thinking. 'I know Benedict has a blob, but he doesn't know that ... Do *I* have a blob? Well, suppose that I *don't* have a blob. Then Benedict will be able to *see* that I don't have a blob, and will immediately deduce from Zeno's remark that *he* must have a blob. But he hasn't shown any sign of embarrassment – oops, that means *I* must have a blob.' At which point he blushes bright red. Benedict does the same, at much the same instant, for much the same reason.

Without Zeno's innocent remark, neither train of thought could have been set in motion, yet Zeno tells them nothing – apparently – that they do not know already.

This effect gets even more puzzling when we try it with *three* monks. Now Brothers Archibald, Benedict, and Cyril are asleep in their cell and Jonah paints a blue blob on each of their heads. Again, when they wake up, each of them notices the blobs on the others, but says nothing. This logical Mexican standoff is broken only when Zeno drops his bombshell: 'At least one of you has a blue blob on his head.'

Well, that starts Archibald thinking, and what he thinks is this. 'Suppose I don't have a blob. Then Benedict sees a blob on Cyril, but nothing on me, and he can ask himself whether *he* has a blob. And he can reason like this: 'If I, Benedict, do not have a blob, then Cyril sees that neither Archibald nor Benedict has a blob, and can deduce immediately that he himself has a blob. Since Cyril, who is an excellent logician, has had plenty of time to work this out but remains unembarrassed, then I, Benedict, must have a blob.' Now, since Benedict is also an excellent logician, and has had plenty of time to work this out but remains unembarrassed, then it follows that in fact I, Archibald, *do* have a blob.' At this point Archibald turns bright red – as do Benedict and Cyril, who have followed closely similar lines of reasoning.

The same kind of argument works with four, five, or more monks – again assuming, for the moment, that all of them have blobs on their heads. Their deductions become more convoluted, but however many monks there are, the announcement that 'at least one of you has a blob' triggers a chain of deduction leading all of them to conclude that they themselves have a blob. When the numbers get large it helps to have some timing device to synchronize their deliberations, and I'll introduce such a device in a moment when we start sorting out what's going on. Similarly paradoxical things happen if some monks have blobs and some don't – I'll come back to that.

There are many puzzles of this kind, involving children with dirty faces, partygoers wearing silly hats, two people who are in possession of consecutive positive integers but don't know who has the bigger one – even a rather non-PC version about marital infidelity among the members of an island tribe. All of these puzzles are distinctly perplexing, inasmuch as the whole procedure is sparked off by somebody announcing a fact that is perfectly evident to everybody. However, when you start to analyse what's going on, it becomes clear that the announcement does indeed convey new information. The informality of language, so often helpful, is in this instance obscuring what goes on.

Let's go back to the first example with the two monks. Zeno announces that 'At least one of you has a blue blob on his head.' What do the monks actually know? Well, Archibald knows that Benedict has a blob, and Benedict knows that Archibald has a blob. But these facts are not the same. When Archibald hears Zeno's statement and concludes that he knew that already, his 'somebody' is Benedict. But when Benedict hears Zeno's statement and concludes that he knew that already, his 'somebody' is Archibald. Not the same statement at all. What Zeno's announcement does is not just to inform Archibald that someone has a blob. It also informs Archibald that Benedict now knows that someone has a blob, and it is the *same* someone. So Zeno's statement doesn't tell Archibald anything new about what Archibald knows, but it does tell Archibald something new about what Benedict knows.

Logical conundrums of this kind are known as 'Common Knowledge' puzzles, and they all rely on the same mechanism. It is not the content of the statement that matters: it is the fact that everybody knows that everybody else knows it. Once that fact has become common knowledge, it becomes possible to reason about other people's responses to it.

Back to the monks. Suppose now that there are 100 monks, each bearing a blob, each unaware of that fact, and each an amazingly rapid logician. To synchronize their thoughts, suppose that the Abbot has a bell. 'Every ten seconds,' says the Abbot, 'I will ring this bell. That will give you plenty of time to carry out the necessary logic. Immediately after I ring, all those of you who can deduce that you have a blob must put your hands up.' He waits ten minutes in silence, save for the repeated ringing of his bell, but nothing happens. 'Oh, yes, silly me, I forgot – here is one extra piece of information. At least one of you has a blob.' Now nothing happens for 99 rings, and then all 100 monks simultaneously raise their hands after the 100th ring.

In essence, the logic goes like this. Monk number 100, say, can see that the other 99 all have blobs. 'If I do not have a blob,' he thinks, 'then the other 99 know this. That takes me out of the reckoning altogether. So they are making whatever series of deductions you get with 99 monks, when I don't have a blob. If I've sorted out the 99-monk logic right, then after 99 rings they will all put up their hands.' He waits for ring 99, and nothing happens. 'Ah, so my assumption is wrong – so I must have a blob.' Ring 100, up goes his hand. Ditto for the other monks.

The 99-monk logic (on the hypothetical basis that monk 100 is blobless) is the same: now monk 99 expects the other 98 to put up their hands at the 98th ring, *unless* monk 99 has a blob. And so on, recursively, until we finally get down to a single hypothetical monk who sees no blobs anywhere, is startled to discover that somebody has one, immediately deduces it must be *him*, and puts his hand up after the first ring.

It's an instance of 'mathematical induction', which says that if some property of whole numbers n holds when $n = 1$, and if its validity for n implies its validity for $n + 1$ no matter what n may be, then it must be valid for *all n*.

So far I've assumed that every monk has a blob, but by similar reasoning you can convince yourselves that this is not an essential requirement. Suppose, for example, that 68 monks out of the 100 total have blobs. Then, with perfect logic, nothing happens until the 68th ring, at which point all those with blobs put up their hands simultaneously, but none of the others.

Common knowledge puzzles have been widely investigated, and some useful references can be found in an article by David Gale (see Further Reading at the end of the book). The most mathematical example described there, and the most far-reaching, was invented by John Conway (Princeton University) and Michael Paterson (University of Warwick, UK). Imagine a Mad Mathematicians' tea party. Each partygoer wears a hat with a number written on it. That number must be greater than or equal to zero, but it need not be a whole number; moreover, some player's number must be non-zero. Arrange the hats so that no player can see their own number, but they can see everybody else's.

Now for the Common Knowledge. Pinned to the wall is a list of numbers. One of them is the total of all the numbers on the players' hats – but nobody knows which is the right total. Finally, assume that the number of possibilities on the list is less than or equal to the number of players.

Every ten seconds, a bell rings, and anyone who knows their number – or equivalently knows the correct total, since they can see everybody else's numbers – *must* announce that fact. Conway and Paterson proved that with perfect logic, eventually some player will make such an announcement.

At first sight, this is paradoxical. Suppose, for instance, that there are three players, and each player's hat bears the number 2, while the list pinned to the wall reads 6, 7, 8. Each player sees a subtotal of 2 + 2 on the other players' hats, so their own hat must be either 2, 3, or 4. Therefore each of the others is looking at either 2 + 2, 2 + 3, or 2 + 4, and any of the totals 6, 7, 8 is possible (remember, some players, though not all, can have zero on their hats). So no total can be ruled out. However, thanks to the bell, players can make inferences from the fact that other players have not yet announced they know their number. At each ring, certain sets of numbers are ruled out, and this leads to Conway and Paterson's unexpected conclusion.

To get an idea of what's involved, consider just two players, and suppose that the list pinned to the wall is 6, 7. The numbers themselves are not known, so call them x and y. What both players know is that $x + y = 6$ or $x + y = 7$. Now for some geometry. The pairs (x, y) that satisfy these two conditions are the coordinates of two line segments in the positive quadrant of the plane (Figure 1a).

If either x or y is greater than 6, then the game terminates after the first ring, because the other player can see immediately that a total of 6 is impossible. The pairs (x, y) for which this happens are shown in Figure 1b. (A little care is needed here: the points (1, 6) and (6, 1), which lie at the ends of the marked segments, are *not* eliminated. The

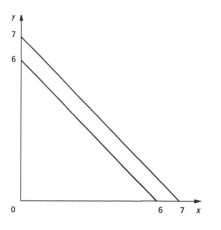

Figure 1(a)

Two line segments correspond to possible numbers on hats.

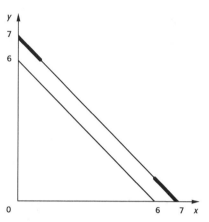

Figure 1(b)

If numbers fall on the segments shown by thick lines, the game ends at the first ring.

eliminated segments are missing one endpoint, the one nearest the middle of the sloping lines.) If neither player responds after the first ring then these possibilities are eliminated. The game then terminates on the second ring if either x or y is less than 1. Why? The other player can see the hat with a number less than 1, and knows that their own number is 6 or less; therefore the total of 7 is ruled out. The pairs for which the game terminates on the second ring are shown in Figure 1c. As this line of reasoning continues, the pairs (x, y) for which the game stops after a given ring form successive diagonals of two 'staircases', one descending from top left and one ascending from bottom right, as shown in Figure 1d. These diagonal segments quickly exhaust the possibilities. In fact here the game must stop by the eighth ring. (Because of

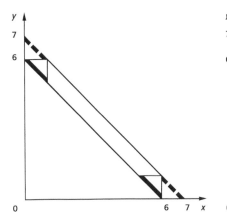

Figure 1(c)

If numbers fall on these segments, the game ends at the second ring.

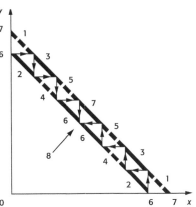

Figure 1(d)

Continuing along two 'staircases' between the lines, we find how long the game continues for any pair of numbers (the required number of rings is marked on the appropriate segments; each segment is missing the endpoint that lies nearest the centre of the sloping lines). Here the biggest number of rings required is 8.

the 'missing endpoints' that I mentioned, the numbers (3, 3) require eight rings. Every other possibility requires seven or fewer.)

The same kind of argument takes care of any two-player list, and even lets us work out the maximum number of rings required. The proof for more players is very simple, but mathematically sophisticated; Gale's article contains full details. As a challenge, work out what happens with three players, each wearing the number 2 on their hat, and the list 6, 7, 8, as mentioned earlier. You should find that nothing happens for 14 rings, and then all three players announce their numbers on the 15th.

2

Domino Theories

No matter how many times you try something and fail, that doesn't prove it's impossible. It just shows that you don't know how to do it. To prove something is impossible, you have to rule out all attempts at a solution. A good way to do this is to find an obstacle that cannot be avoided – an 'invariant'. Sometimes you can find an invariant by introducing a few colours, and counting.

Business, the stone-carver Rockchopper Rocknuttersson pointed out to his apprentice, was bad.

'You can say that again,' said Pnerd.

'Business is bad, Pnerd.' The President of the Obelisker's Guild tended to take everything literally. 'If we don't get a commission soon, I'll have to hang up my chisel and take that job herding pigs that my uncle Hogthumper Hogtrottersson keeps trying to offer me.'

Pnerd chipped idly at a child's toy dolmen. 'It's the recession, Rocky. Nobody's buying. The market in stone circles has hit rock bottom. And as for longbarrows ... you can't even sell a *wheel*barrow at the moment. I heard Moloch Molochsson complaining the other day because the tithes are down again and the priests can barely afford to buy enough rams to placate M'gaskil the snow-god before Winter comes.'

Rocky scratched his ample nose, flexing his huge biceps. 'Did you pick up that copy of *Rolling Stone* that I asked you to buy?' Pnerd dropped a large round slab of limestone at his feet. Rocknuttersson picked it up and pored over the chiselled inscriptions. 'There may be something in the small ads. Mmmm ... assistant newt-trainer ... the Bogtown chief vermin inspector has retired ... seven virgins wanted for unspecified purposes, must be willing to travel ... Ah! Invitation to tender for repairs to the marketplace in Quagville! Pnerd, get over there and find out what they want done, while I check the thongs on the dressing-mallets.'

Two days later, Pnerd returned.

'Well?'

'Quagville marketplace is paved with big stone slabs, Rocky. Sixty-four of 'em, each one about ten feet square, arranged in an eight by eight grid. The original stone is starting to crack. They want the whole lot ripped out and relaid.'

'Brilliant!'

'Wait, there are some conditions. The main one is, they don't want it relaid in squares this time. The town priests reckon that's what caused the stone to crack.'

'Rubbish! Typical priests, always worried about shapes and numbers and soft intellectual numerosophist trash ... I know exactly what happened. When Chalkhacker Chalkwhackersson laid those slabs, he used inferior quality stone, and the frost got in.'

'The priests say they cracked because a square is the symbol of Frozo the frost-demon.'

Rocky looked up in surprise. 'Is it? I thought it was the sigil of Gnashfang the cave ogre.'

'That too,' admitted Pnerd. 'But there aren't that many symbols to go round, you know. The square is kind of popular. Gnashfang shares it with Frozo; he gets to use it on alternate weekdays.'

'Oh.' Rocky thought for a few moments. 'Maybe the priests are right, then.'

'Depends whether the frost comes on a Tuesday or not. But right or not, you don't argue with priests. Not if you want to keep your kidneys. Square slabs are out. They want dominoes.'

Rocknuttersson stared at him as one might at something slimy that has crawled from under a rock. 'Pnerd, what in the name of the Great Boggie is a *domino*?'

'Two squares stuck together, Rocky.'

'Then why not say so? Why not make it clear that they want tween-twines? Why use a silly name like 'domino'?'

'Dom'd if I know,' said Pnerd, and dodged the kick that Rocky aimed at him. Then his face fell. 'Could be a problem, Rocky. Maybe dominoes won't fit.'

'Of course they'll fit! All you have to do is place one of them where two of the old squares used to go!'

Pnerd frowned. 'Yes, but that only works if the total number of squares is even. Each domino covers *two* squares. If you started with an odd number, there'd be one square left over at the end.'

Rocky sighed. 'Pnerd, you said there were 64 squares! That *is* even!'

'Is it?'

'Provided the slabs were laid horizontally, yes, the entire *marketplace* must be even.'

'Oh. Right.' He scratched his nose distractedly. 'Um. I guess I should have mentioned the statues of Gog and Magog.'

Rocknuttersson leaped to his feet in anger. 'Statues? *What* statues?'

'The ones I forgot to mention. Seems that when the first slab cracked up, the priests tried to cover up the problem by installing a statue of Gog instead. Soon after, another slab cracked, so they put in a matching statue of Magog. Each one has a base just the same size and shape as one of the square slabs. So it's not 64 squares any more, it's – um ...'

'Sixty-two.'

'Right, yeah. Um – is *that* an even number?'

Rocky began counting on his fingers, but they ran out before he got far enough. 'To be quite honest, Pnerd, I have no idea.'

'Well, you'd better be certain before we chisel our name-signs into any binding legal agreements, Rocky. There's penalty clauses.' He waited for 20 minutes while Rocknuttersson cursed whichever son of a dog had invented penalty clauses in local government contracts, learning 73 new swear words in the process. 'Ten years in the sulphur mines if the new slabs don't fit,' he added by way of explanation. The curses renewed. Eventually Rocknuttersson stopped to draw breath, and Pnerd grabbed his chance. 'Rocky, we can't work this out on our own. We need an expert.'

'Who do you have in mind?'

'Snitchswisher!'

'May the gods preserve you against possession by demons.'

'No, I didn't sneeze, you idiot! Snitchswisher Wishsnitchersdorter!'

'There, you've done it ag – oh, her! Your numerosophist friend who lives in Dead Cat Swamp.' Pnerd nodded. 'Smart thinking, apprentice: we definitely need an expert. We are out of our depth on these matters.'

Snitchswisher Wishsnitchersdorter was sewing new moletail trimmings on her tunic when they arrived. Rocky explained their problem, and she gave a sardonic laugh. 'Good job you came to me. There are aspects of the matter that would not be apparent to lay persons, and you might have got yourselves into serious trouble. To begin with, although 62 is indeed an even number – ' she paused while Rocky and Pnerd argued over who had first conjectured this to be true, and who had claimed the opposite – 'it is not enough for the number of squares to be even.'

'It's not?'

'No. There is a more subtle question of parity. It is an old numerosophical chestnut. For example, suppose that two opposite corners of the square are removed (Figure 2a). Is it possible to cover the remaining 62 squares with dominoes?'

'Beats me,' said Rocky.

'Should be,' said Pnerd. 'There's lots of room to try different arrangements, and there can't be one left over.'

'True. But there might be two left over.' Snitchswisher rummaged in a corner of the hut, and found a board marked off into a grid of 64 squares, and a box of wooden rectangles, each just large enough to cover two adjacent squares. She placed a pebble on the two opposite corners. 'Try it.'

Pnerd began playing with the squares. Rocknuttersson sidled up closer to Snitchswisher and asked what the board and wooden pieces were for. 'I had an idea for a game,' she said. 'The board represents a river, and you have to use the wooden pieces to build a kind of arch thing over it, without it collapsing. I was going to call it 'bridge'.'

'Never catch on, not with a name like that,' said Rocky.

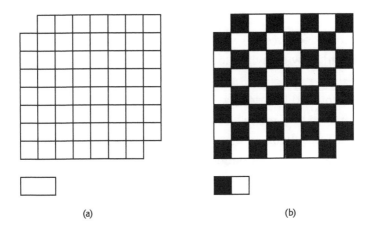

(a) (b)

Figure 2

(a) An 8 × 8 grid with opposite corners removed. Can it be covered by 31 dominoes?

(b) If the grid is coloured like a checkerboard, there are 32 black squares and 30 white. But each domino must cover one of each. Therefore two black squares must remain uncovered.

Pnerd thumped the table in frustration. 'They won't fit! I've tried a dozen times but they won't fit!'

Snitchswisher Wishsnitchersdorter smiled. 'And they never will, Pnerd. Let me draw your attention to the different colours of the squares (Figure 2b).'

'That's a pretty pattern.'

'Yes, I call it 'check'.'

'Why?'

'Because when you draw it you have to be careful to check you haven't made a mistake. I did the black squares with charcoal and the white with extract of arrowroot steeped in deadly nightshade.'

'Why not use chalk?'

'That's a brilliant idea, Pnerd! It's never occurred to me that you can use chalk to write with. Imagine, writing with a rock instead of a burnt stick! Anyway, if you think about placing a domino on the board, you'll see that it must always cover one black square and one white, because no two black squares are adjacent, and the same is true of the white ones. Pnerd: how many white squares are there – not counting the two corners?'

Pnerd counted laboriously. 'Thirty.'

'Right. And how many black?'

'Um ... Thirty-two.'

'Precisely. Since any domino covers one of each, at least two black squares must remain uncovered. You're right that you won't have just one square left over. But that doesn't rule out having *two* left over! It's a general parity principle for dominoes: as well as the total number being even, you also have to have equal numbers of black and white squares.'

'That,' declared Rocknuttersson, 'is absolutely brilliant, Snitch-swisher. Except,' he added, 'that the squares in Quagville market are *all the same colour*.' He gave her a withering look. 'Typical ruddy theorist, no practical sense at all.'

'But,' said Snitchswisher, 'you can always *imagine* the squares are coloured, and the same argument applies.' Rocknuttersson thought about this for a few minutes, and then turned bright red. To cover his embarrassment, he dispatched Pnerd back to Quagville to check that the statues of Gog and Magog had not been placed on squares that, if you *imagined* the marketplace coloured in a black-and-white check, had the same colour.

Two more days passed, during which time Rocky helped Snitch-swisher Wishsnitchersdorter make enough bognettle soup to carry her and her aged father through the coming winter. Then Pnerd reappeared.

'Bog, that was boring. I wrote a poem along the way to amuse myself, Snitchswisher. Would you like to hear it? It's about a fearsome creature of the forest.'

'Proceed.'

Pnerd drew in a breath and poked his skinny chest out. 'Rabbit,

rabbit! Burning bright, in the woodlands of the night. What immortal hand or eye – '

'Could turn you into rabbit pie,' said Rocky. 'Stop wasting time, Pnerd, and report on the placement of the statues.'

'We're in business, Rocky! One statue is on a white square, one on a black one!'

'Which?'

'Eh?'

'Is Gog on black, or white?'

'Crumbs, Rocky – '

'Look, it could be important, Pnerd. The priests of Gog wear black cloaks, whereas those of Magog wear – '

'Oh. Look, they were only imaginary colours anyway, Rocky, I could always change them roun – '

Rocky suddenly shook his head. 'It's not that easy, Pnerd. I've just realized that the priests of Magog wear black hats, whereas those of Gog – '

'For Bog's sake!' yelled Snitchswisher. 'Who *cares*?' She grabbed Pnerd by the shoulder. 'You don't happen to recall just where the two statues *were*, do you?'

'Nope.'

'Oh, heck.'

'Does it *matter*?' asked Rocky.

'I'm not sure. It might. Should we send Pnerd back to – no, that'll take days.'

'*Two* days,' said Pnerd. 'And I'm fed up flogging it to Quagville market, anyway.'

Snitchswisher looked thoughtful. 'You know, maybe it doesn't matter,' she said. 'But you'd have to try an awful lot of possibilities to be sure. I think it's time we consulted my father.'

'Her dad's a thaumaturge,' Pnerd reminded Rocky. 'Gets in touch with spirits, that sort of thing.' Rocky seemed sceptical, possibly because he always ended up out of pocket when thaumaturgy was involved, but Snitchswisher trotted off through the bog to fetch him. Soon she and the old man – Wishsnitcher Dishpitchersson by name – reappeared. Suitably primed with silver from Rocknuttersson's purse, he fished some Tarot cards from his robes and began a divination.

'Below ... the Moon. Above ... the Leaping Cow. To west and east ...'

'The Cat and the Fiddle,' suggested Rocky.

'Yes, but the Cat is inverted, signifying drunkenness ... Below, the Laughing Hound – '

'Signifying that these entire proceedings are a farce – '

'Signifying merriment. More cards ... the Dish, the Spoon – '

'And the Knife and Fork.'

'No ... the deuce of Forks.' The old man shook his head. 'Which is strange, since there are no forks in the pack ... Ah! A name ... A spirit from the future ... an acolyte of 'Big Blue', whatever mystic being that may be ... Ralph ... Ralph ...'

'That's the hound. They always go 'Ralph! Ralph!' But it's called 'barking', not 'laughing'.'

'No, it is a name ... Ralph ... er ... Grimoire? Grimory? No – Ralph Gomory, a future numerosophist of great ingenuity ... A three-tined fork and a four-tined fork, a sigil of enormous power and beauty. Quick, the charcoal!' The old man drew swift lines on the board (Figure 3). Then his trance faded.

Rocknuttersson sourly handed over a further piece of silver. 'I reckon your dad's a few stones short of a henge, Snitchswisher.'

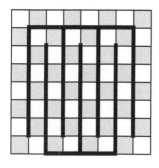

Figure 3

Gomory's sigil creates a chain that can be filled by consecutive dominoes.

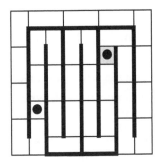

Figure 4

How to fill in the dominoes if the two omitted squares are of opposite colours.

She sniffed and studied the charcoal lines. 'I am not so sure, Rockchopper Rocknuttersson. Imagine the two forks are walls. Then a line of dominoes may be placed between them, in an endless loop. If two squares are occupied by statues, the loop is cut into two sections. Possibly just one if the squares are adjacent. If the statues are on squares of the opposite colour, then each section contains an even number of squares, so the chain of dominoes can fill it completely. The diagram represents a proof that no matter which two squares are occupied by statues – provided only that they are of opposite colour – the remainder can be covered by dominoes. Indeed it is a constructive proof, showing exactly how to achieve such a result in any given case' (Figure 4).

Rocky was impressed. 'Snitchswisher, I apologize to your father for my scepticism. He has uncovered a remarkable truth.' The old man muttered something about 'fine words', 'butter', and 'parsnips', and Rocky handed over another piece of silver to avoid further embarrassment. 'Pnerd! Fetch my scribing-chisel, and the finest slab that is portable! We shall head the document TENDER FOR RENOVATION OF QUAGVILLE MARKETPLACE, ROCKHOPPER ROCKNUTTERSSON ROCK RENOVATORS, MURKLE MIRE.'

'OK,' said Pnerd. 'Oops, I never told you about the two new statues, did I?'

Rocky stared at him. 'Two ... new ...'

'Demagog and Wolligog. The priests decided to cover up some more cracks.'

'Oh my Gog,' said Rocky.

'They are on squares of different colours,' said Pnerd helpfully. 'Two statues on black squares, two on white.'

'You don't happen to remember exactly wh – no, of course you don't. Snitchswisher: does the sigil of Gomory work when there are four missing squares, two of each colour?'

Snitchswisher Wishsnitchersdorter's brow furrowed. 'It does if the order in which the squares appear as one passes round the loop of dominoes is alternately black and white,' she said. 'But if black is followed by black, then the number of intervening squares is odd, and the proof breaks down.'

'And could that happen?'

'I don't see why not. It's all rather confusing.'

'You're right about that.' There was a lengthy pause. Rocky started to say something, but was interrupted.

'No, quiet! I'm getting an idea ... Yes, of course. Cut the board into two pieces, such that each contains just one missing square of each colour. Do this so that each piece can be covered by an endless loop of dominoes, like Gomory's sigil but of whatever shape suffices. Then use the same argument to prove that each piece can be covered.'

'Are there such sigil-bearing pieces?'

She thought for a moment. 'Many. I shall draw some' (See Figure 5).

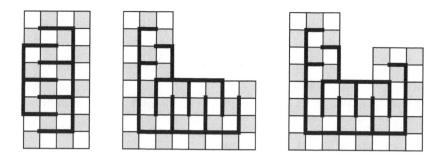

Figure 5 Some sigil-bearing regions.

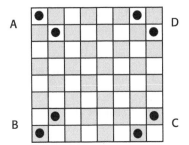 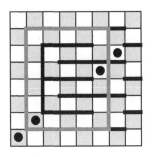

Figure 6

Four problematic corner arrangements. C and D block the placement of a domino to cover the corner square, but A and B are harmless.

Figure 7

An example of how to handle case A of Figure 6. Each sigil-bearing region contains precisely one omitted square of each colour.

'Hmmm ... I haven't time to enter into every detail; but I'm pretty sure you can show that the only occasions on which the board cannot be so divided are when either the two black squares omitted, or the two white squares, are in the same corner (Figure 6). In one arrangement, it is obvious that the corner square is isolated from all the rest, and no solution can be found. In the other case ... the board can again be divided into two regions, each containing just one omitted square of each colour, and each possessing a Gomory sigil of its own (Figure 7). One region has to have a hole in it, but that doesn't alter the argument. I believe that a careful analysis will show that it is always possible to cover the board with dominoes, except when a corner configuration such as in Figure 6 occurs, for one or other colour.' She shrugged. 'It is not as elegant a proof as Gomory's, however. Perhaps some future numerosophist can do better.'

'Anyway,' said Rocky, 'it sounds as if we're probably in business.' He leaped to his feet. 'What we need is somebody to go and check that the statues haven't cut off one of the corner squares. To be on the safe side, Pnerd, this time you can also make a map of the positions of the statues, so that we know exactly what we're up against. Then we can use Snitchswisher's wooden pieces to find a solution *before* putting in our bid to tender.'

Pnerd groaned. 'Crumbs, why me? I've been twice already and it's a two-day walk every – '

'*You*, Pnerd, are the apprentice. *I* am President of the Obelisker's Guild.'

'Oh, right. I'll get started then.' He borrowed a few strips of candied goat to eat on the journey, and headed for the door.

'Oh, and Pnerd?'

'Yes, Rocky?'

'It would be nice if you could get back before the priests put up any *more* statues.'

3

Turning the Tables

When it's time to rearrange the furniture, but space is limited, the order in which you move things can make a big difference. But how can you find the right order, and the right moves? To find your way through a city, or a maze, it helps to have a map. So what you need is a map of the puzzle — a conceptual map of a logical maze.

Sixty-seven floors up in the Ruff Tower, two employees of We-Haulit-4U Moving Company struggled with the last of nine solid oak tables, every one of which had been carried by hand up the narrow twisting staircase that was normally reserved as an emergency fire exit. They would have used the elevator but Goofy Ruff, the owner of Ruff Towers, was worried that the tables' weight might be too much for the suspension cables.

They dragged the table into the storeroom to join the other eight. The door clicked shut behind them.

'Done,' said Dan, breathing heavily. 'Final check, and then I'll buy us lunch at the Plushy Pink Pizza Palace. Two jumbo-sized square tables, six humongous rectangular tables, and one top-of-the-range megabronto.'

'Right,' said Max, ticking off the items on a scruffy clipboard. 'One one by one, six two by ones, and one two by two.' His pen scribbled across the page. He looked up. 'Say, it's a bit crowded in here.'

'Jam packed. Wall-to-wall tables, except where we're standing.'

'We did well to fit them in. I wonder why they want so many?'

'I think they're just using this room for temporary storage until the redecoration of the ballroom on the ground floor is finished. Word is that Rasputina Ruff told Goofy she really preferred lime green to turquoi – '

Max groaned. 'You mean we carried this stuff all the way up here, and they're going to want it back down there?'

'Yup. Next week. It's business, Max, don't knock it. Think of it as a challenge, a test of mental character and physical strength. I can't resist a challenge, can you?'

'I have enough mental character and physical strength to resist *any* challenge, thank you very much. I think I'll get a job digging drains, it's nearer the ground.'

'Speaking of which, so is the Plushy Pink Pizza Palace.'

'Right. Let's go. Woops.'

'Whaddya mean, woops?'

'The door must have locked itself behind us.'

Dan's face was a picture. He gathered his wits. 'No need to panic, there's supposed to be an emergency phone somewhere.'

'I know,' said Max. 'It's behind that small door in the wall marked "emergency phone".'

'Great.'

'Which is blocked by that solid oak table.'

'Not so great. We'll have to move it.'

'They're packed pretty tight,' Max observed. 'It's not going to be easy.'

'Can't we pile 'em up somehow, make a bit of space?'

'Not a chance. Ceiling's too low.'

After half an hour of futile effort, they called a halt. 'Dan, we've got to think this one through before we run out of energy. I reckon we'd be OK provided we could move the megabronto table into that corner over there (Figure 8). We can slide the tables into the space that's left, one by one, thereby creating new spaces to slide more of them into.'

'Won't we get trapped?'

'No, we can crawl underneath,' said Max.

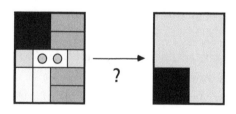

Figure 8

Can you slide the tables until the big black square moves to the lower left corner? Dan and Max, marked by two circles, are standing in the only open space.

Dan stooped to peer under a table. 'You're right, there's plenty of room.' He scratched his head, thinking. 'You know,' he said, 'when I was a kid I used to have a toy. It was called 'Dad's Puzzler', and you had to slide rectangular and square blocks around so that Dad could move his piano. It was quite like this.' He paused. 'Suspiciously like this, in fact. Anyway, it took me a while, but in the end I learned how to solve it.'

'Great! Can you remember how?'

'Yeah. You slide the blocks around until you get them where you want.'

Max grimaced. 'I think we need something a little bit more specific, Dan.'

Dan shrugged, it wasn't his fault that he couldn't remember how to solve a puzzle he'd been given for his sixth birthday. 'I can still recite the whole of *The Cat in the Hat*,' he said, by way of demonstrating that he was nevertheless a possessor of a super-power memory.

'Yeah. "And all we could do was to sit sit sit sit. And we did not like it, not one little bit." Thanks a bunch, Dan.'

'No use moping. Let's shift a few of the tables and see where it gets us.'

Another half hour passed, after which they had successfully moved the megabronto table from the top left corner to the middle of the right-hand wall (Figure 9). It was progress, but – as Dan remarked – in which direction?

'What we need,' mused Max, 'is a map.'

'Max, we can *see* where all the tables are.'

'Not a map of the *room*.'

'What, then?'

'A map of the *puzzle*.'

Dan stared at him. Have you gone crazy? Puzzles don't have maps.'

'I hate to contradict you, old buddy – no, come to think of it I like to contradict you, but anyway – puzzles do have maps. Conceptual maps. Imaginary maps in the brain. Maps that tell you what all the positions

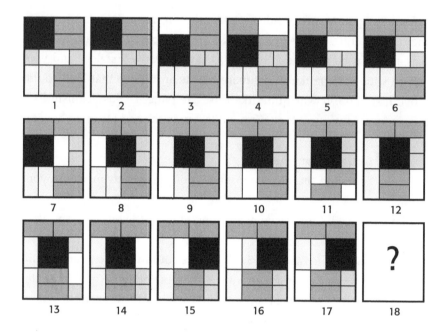

Figure 9 A possible sequence of moves. There's a long way to go.

in the puzzle are, and how to get from one to the other. Mental mazes that tell you what moves to make, and in what order.'

Dan nodded. Of course. Except ... 'It's going to be a pretty complicated map, Max. There's an awful lot of positions, and an awful lot of moves.'

'True. So we'd better find some way to cut them down. Break the problem into simpler pieces. Hey! Yes, that's it. First of all, let's find out what we can do *easily*. Then we can try to kind of string those together.'

'Well, for a start, if you've got a square hole with just the two smallest tables in it, you can move those tables around pretty freely,' said Dan. See Figure 10a.

'Yeah, that's the kind of idea. A sort of 'sub-puzzle' where you move only a few tables around, inside some well-defined boundary.' (Positions 5-6-7 of Figure 9 use just such a sub-puzzle.) He stopped and

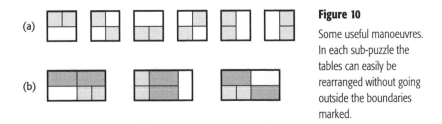

Figure 10

Some useful manoeuvres. In each sub-puzzle the tables can easily be rearranged without going outside the boundaries marked.

thought. 'Hmm. There's another, a bit more complicated, where you have a rectangular region containing two rectangular tables, two square ones, and the rest free space.' See Figure 10b.

'So you could assume that positions that differ from each other by shuffling tables around inside one of these sub-puzzles are effectively the same,' said Dan. 'That must cut the list of positions down quite a bit.'

'Yeah. And there's another thing. Sometimes there's only one sensible way to continue moving the tables, if you don't want to just go back on yourself.' (Positions 3-4-5 in Figure 9, or the longer sequence 7–17, are examples.)

'So provided you know where you're starting from and where you're trying to go, sequences like that can be left off the map?'

'Precisely. Hand me that clipboard and pen.' Shortly, Dan and Max were staring at a map of part of the mental maze of possible positions and moves (Figure 11).

'I've marked the start and the finish positions,' said Max. 'Then there are various ways to place key tables, which I've marked A, B, C, D, E, F.'

'I'd have expected there to be more than just six of those.'

'There are. This is just *part* of the map. But it's more than enough to solve the puzzle. Now, shut up and listen. The lines show sequences of forced moves – in the sense that if you know where to start and where to finish, the moves in between are fairly obvious because on the whole there's only one choice you can make at each step, right?'

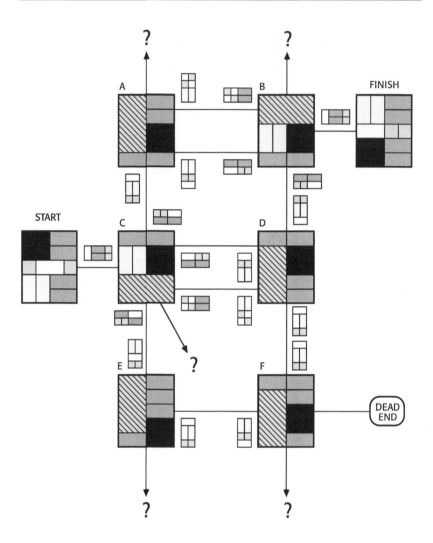

Figure 11

A partial map of the puzzle. Large diagrams show the placing of key tables.
Cross-hatched regions represent sub-puzzles to be solved using the useful manoeuvres of
Figure 10. Lines indicate sequences of moves, sometimes quite long, that are essentially
'forced' if you know where you want to go. For example the line from START to C represents
the sequence of 17 moves shown in Figure 9. The small diagrams show how the sub-puzzles
are to be arranged at the start and finish of these sequences. Using this map as a guide,
the puzzle becomes relatively easy to solve.

'OK, I see that. Once you've played around with a puzzle for a while, you can't help noticing that kind of thing.'

'You said it. Now, I've shaded in rectangular regions where there's a sub-puzzle to solve. To show *which* sub-puzzle, I've drawn little pictures of the start and finish positions within the rectangle, at the appropriate ends of the connecting lines.'

Dan's mouth opened like a goldfish's. 'Sorry, I don't quite follow.'

'Well, suppose you want to work out how to move from C to E. Look at the vertical line that joins them. Beside it are two little diagrams. If you replace the cross-hatched area in C with the top diagram, and the cross-hatched area in E with the bottom one, that gives you the start and finish positions. Because the moves in between are 'forced', it doesn't take very long to work them out. If you make a copy of the puzzle out of bits of card, you can move them around and check.'

'What does DEAD END mean?'

'What do you *think* it means? Now, what does the map tell us?'

'Where things are and how to go between them. Well, clues to those things.'

'It tells us more than that. It tells us that one way to solve the puzzle is to go along the route START-C-A-B-FINISH. Just use the little diagrams beside the appropriate lines to fill in the shaded bits of the big diagrams, then follow the forced sequences of moves.'

Dan's face lit up in admiration. 'You could go START-C-D-B-FINISH instead?'

'Sure. Or even START-C-E-F-D-B-FINISH – but that would be an unnecessarily complicated route.'

Dan was getting into it now. 'Or START-C-D-F-E-C-D-B-A-B-D-C-E – '

Max interrupted before his friend collapsed from lack of breath. 'Yeah. But that would be an even *more* unnecessarily complicated route.'

'I'll settle for the simplest one.'

'Fine by me. Let's get these tables moving!'

It took a while to get into the swing of it, but once they did, it didn't take long to get the megabronto table into the lower left corner of the room. After that, Max was able to get his hands on the emergency phone and call the porter in the lobby. When help arrived it turned out that the new arrangement of tables was blocking the door so it wouldn't open, but by now Dan and Max knew their way around the map of Dad's Puzzler blindfolded.

Not long after midnight, they were free.

Somewhat shaken by the experience, they flagged down a cab and set off for the Plushy Pink Pizza Palace, which was an all-nighter. First they'd catch up on lunch, then they'd have dinner.

'You know,' said Dan, 'that wasn't so hard.'

'Not once we'd worked out that map. But we were lucky, it was quite a simple one.'

'Yeah. But that's because you used some tricks to simplify it.'

Max rubbed his chin – to find a heavy growth of stubble. 'The tricks help, but there are plenty of sliding block puzzles with much more complicated maps, even when you use every trick you can think of.'

'Like what?'

'Well, there's one called the Donkey puzzle, which is probably from the nineteenth century and almost certainly French. That's a good bit harder. The Century puzzle, invented around 1980, is harder still. You must make 100 moves to solve it. And if you insist that the finishing position should be like the start but upside down, then it's *really* hard. That version is called the Century and a Half puzzle because it requires 151 moves.' (See Figure 12.)

The cab screeched to a halt outside the Pizza Palace; Dan paid the driver. They went in and sat down. Max ordered a deep pan pizza with extra cheese. Dan ordered a special with a lot of extra toppings – pepperoni, tuna, capers, anchovies, beef, pineapple, hot tamales, a whole banana, chewing-gum, liquorice, and a lighted sparkler. 'My favourite,' he explained to a bemused waitress. 'Make sure you build it from the bottom up in the order I specified.'

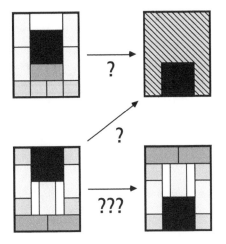

Figure 12

Three harder sliding block puzzles. Any arrangement of blocks is allowed in the cross-hatched region.

Top arrow: the Donkey puzzle.

Middle Arrow: the Century puzzle.

Bottom arrow: the Century and a Half puzzle.

The pizzas arrived. Dan's didn't look quite right. Most of the ingredients were upside down, including the crust. The waitress had included a whole tuna and tried to set the liquorice on fire.

'Enjoy your puzzle, sir,' she said over her shoulder.

'Send it back,' suggested Max.

'No, no, you heard what she said. I can't resist a challenge, it's a test of character. The pizza just needs rearranging.' Dan picked up the tuna, tried to find somewhere to put it while he blew out the liquorice. Where had the pepperoni gone? Oh, yeah, inside the pineapple. The plate just wasn't big enough ... He sighed, put the tuna back, and was on the point of calling the manager to complain that his pizza was too difficult to solve.

Then straightened his back, squared his shoulders, and reached for the clipboard.

'What are you doing?' asked Max.

'I can fix it. Just wait till I've made a map of this pizza.'

4

The Anthropomurphic Principle

Murphy's Law says that if something can go wrong, it will go wrong. And if something can't possibly go wrong, it will go wrong anyway. For instance, if buttered toast falls off the table, it will always land buttered side down. (Unless you buttered the wrong side ...) But is that really a case of Murphy's Law, or might it be an unavoidable property of the physical universe?

I've never had a piece of toast
Particularly long and wide,
But fell upon the sanded floor,
And always on the buttered side.

S**o wrote the poet James Payn** in a parody of Thomas Moore's lines about a gazelle in *The Fire Worshippers*. The event described is the archetypal instance of Murphy's Law: 'if anything can go wrong then it will'. Its origins lie in experiments carried out in the late 1940s by a USAF Captain (no prizes for guessing his surname). It has many variations and extensions, such as 'even if it can't possibly go wrong, it still will,' and it appears under other names than Murphy's.

In 1991 the British Broadcasting Corporation's TV series *QED* carried out experiments in which people tossed toast into the air under various conditions, and in every case the results were statistically indistinguishable from pure chance. And there the matter might have rested, were it not for Robert Matthews. Matthews is a British journalist with a mathematical streak: a typical Matthews calculation starts with, say, a photograph of a building whose windows have blown out, and ends with an estimate of the wind speed. In the *European Journal of Physics* (see Further Reading) he observes that there are two problems with *QED*'s experiments. First, by its nature, Murphy's Law may conspire to falsify any experiments aimed at testing it. Second, in the normal circumstances of eating breakfast, toast is not hurled randomly into the air. (OK, *your* family may have its own way of doing things, but my basic point stands.) Toast is generally knocked sideways off the edge of a table, and any experiment ought to build this fundamental feature into the experimental design and analysis.

Before proceeding, it is worth exposing one common misconception. The asymmetric behaviour of falling toast is *not* the result of the extra mass of butter. A typical piece of toast weighs about 40 grams, the butter is at most 10% of the total, and anyway it is mostly absorbed into the central regions. Its effect on the dynamics of the toast is negligible. And its effect on the aerodynamics of the toast, resulting from changes in surface viscosity, is even more negligible.

Matthews traces Murphy's Law to a much simpler asymmetry: it is the top surface of toast that is buttered, and that surface remains on top when the toast is first nudged over the edge of the table. As the toast falls towards the ground, it rotates at an angular velocity determined by the degree of initial overhang of its centre of mass. Might it be the case that the height of a normal table and the Earth's gravitational force conspire to create a predominance of rotations through an *odd* multiple of 180°? If so, it will land butter-side down every time. And the short answer is that, according to Matthews's calculations, they do. In fact a rotation that flips the toast over just once, leading to a butter-down final state, is by far the commonest.

Before we consider the deeper reasons for this unhappy coincidence, it is worth summarizing the mathematical arguments that lead to this conclusion. Figure 13 shows the initial configuration of the toast and the main variables involved, together with some key formulas derived from Newton's laws of motion. The main conclusion is that the toast cannot land butter-up unless the 'critical overhang parameter' – the percentage of toast that hangs over the edge initially, relative to half the width of the toast – is at least 6%. Experiment shows that for bread the value is 2%, and for toast 1.5%. Both fall far short of what is required to make the bread or toast rotate through at least 360° on its way to the floor. Since the rotation is provably at least 180°, this implies that butter-down is the inviolable rule.

THE MURPHODYNAMICS OF TOAST

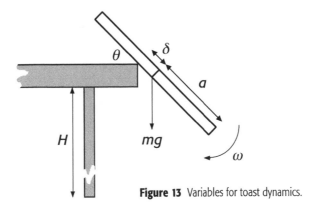

Figure 13 Variables for toast dynamics.

The key variables (see Figure 13) are:

g = acceleration due to gravity

m = mass of the toast

a = half-width of the toast

δ = initial overhang

θ = angle of rotation

ω = angular velocity of rotation

H = height of the table.

Define the *overhang parameter* $\eta = \delta / a$. Then Newton's laws of motion lead to the relation

$$\omega^2 = (6g/a)(\eta /(1 + 3\,\eta^2))\sin\theta$$

as long as the toast is pivoting about the edge of the table. The toast begins to slip when the frictional force at the table's edge is exceeded by the appropriate component of the toast's weight. Whatever the rotation rate is at that instant, the toast will thereafter rotate at the same rate during its fall.

Simple estimates show that the toast will flip through at least 180° on the way to the floor. In order to land butter side up, it must therefore rotate at least 360°. We know how fast the toast is rotating, and H (together with g) tells us how long it takes to hit the floor. For tables and toast of conventional dimensions, Matthews shows that it rotates at least 360° only when the critical overhang parameter h is greater than 0.06. The critical overhang occurs when the toast first detaches itself and begins to fall freely.

This analysis makes a number of assumptions. One is that the toast does not bounce when it hits. Since butter is highly viscous, this is reasonable: the normal outcome is a *splat*, not a *boing*. Another is that the toast slides *slowly* over the edge so that it detaches at the critical overhang value. A more detailed analysis shows that the horizontal velocity imparted to the toast as it goes off the edge of the table has no serious effect on the conclusions unless it is at least 1.6 m/sec, a fairly hefty wallop. This result does lead to a strategy for preventing butter-down descent: if you notice the toast sliding off the edge, bat it firmly with your hand to send it skimming across the room. This strategy is quite likely to have other adverse effects – for example if the family cat is sitting at the point of impact – but it should avoid buttering the carpet.

This analysis is all very well, but it suggests that Murphy's Law is merely a coincidence, a strange case of 'murphic resonance' resulting from the rather arbitrary values that human culture assigns to tables and toast, in conjunction with the equally arbitrary value of the Earth's gravitational field. Matthews goes on to observe that nothing could be further from the truth. Murphy's Law, as embodied in twirling toast, is a deep consequence of the fundamental constants of nature. Any universe that contains creatures remotely like us will necessarily inflict Murphy's Law upon those creatures – at least if they eat toast and sit at tables.

The precise argument is technical and convoluted, but its outlines are simple. The key fact was stated by W. H. Press, who argued in 1980 that the height of a bipedal organism is limited by the gravitational field in which it lives. Compared to quadrupeds, bipeds are intrinsically unstable: they are far more likely to topple because their centre of mass only has to stray outside their 'footprint' for them to fall over. Quadrupeds have a much larger region of stability. (It is no coincidence that giraffes are taller than humans.)

The critical height is the one for which the impact of the head with the ground is likely to cause death. Of course this argument assumes that the crucial equipment is located at the top of the biped, but this does offer evolutionary advantages such as the ability to see further. Half the fun in this kind of discussion is to make plausible assumptions and see where they lead. The other half, which I leave for you to think about, is to deny those assumptions and see where *that* leads.

It is also reasonable to assume that the height of a table used by an intelligent biped will be about half the creature's own height. On Earth, a table needs to be about 3 metres high in order for Murphy's Law to be violated, so we would have to be 6 metres high to escape the unfortunate consequences of murphic resonance. The deeper question is: might some race of aliens on some distant planet be murphologically immune?

To answer this question, Matthews models the said alien as a cylinder of polymer whose critical component is a sphere positioned at the top. I will call such an organism a polymurph. Death is occasioned by the shearing of chemical bonds in a polymer layer. His analysis leads to the conclusion that the height of a viable polymurph is at most

$$(3nq/f)^{1/2}\mu^2 A^{-1/6}(\alpha/\alpha_G)^{1/4}a_0$$

where

n = the number of atoms in a plane across which any breakage takes place (typically about 100)

q = 3×10^{-3} is a constant related to polymers

f = the fraction of kinetic energy that goes into breaking the polymer bonds

μ = the radius of polymeric atoms in units of the Bohr radius

A = the atomic mass of polymer material

α = the electronic fine structure constant $e^2/(2h\varepsilon_0 c)$ where e is the charge on the electron, h is Planck's constant, ε_0 is the permittivity of free space, and c is the speed of light

α_G = the gravitational fine structure constant $2Gm_P^2/hc$ where G is the gravitational constant and m_P is the mass of the proton,

a_0 = the Bohr radius.

Plugging in the relevant values for our universe we find that the maximum safe height for a polymurph is 3 metres. (The tallest recorded human being, by the way, is a certain Robert Wadlow, height 2.72 metres.) This is far short of the 6 metres needed to avoid buttering the kitchen carpet.

Interestingly, this upper limit on polymurph height does not depend upon which planet the alien inhabits. The reason is that the balance between internal gravitational forces and electrostatic and electron degeneracy effects, required for the polymurph not to fall apart, relates the planet's gravity to more fundamental constants. Thus we find that Murphy's Law is not a coincidence at all, but the consequence of a deep 'anthropomurphic principle': any universe built along conventional lines that contains intelligent polymurphs will conform to Murphy's Law. Matthews concludes 'According to Einstein, God is subtle but He is not malicious. That may be so, but His influence on falling toast clearly leaves much to be desired.'

My original column ended there, but it attracted such an unprecedented level of comment, including alternative histories of the origin of the phrase 'Murphy's Law' and objections to its use to describe inanimate behaviour, that it is worth recording a few of the responses from readers. David Carson of the Mississippi University for Women reported a series of experiments conducted by a group of faculty and students, in which toast was (a) tossed randomly from waist height, (b) pushed off the edge of a table, and (c) pushed off a ten-foot aluminium ladder. In cases (a) and (c) the observed frequency of toast landing buttered side down was 47% and 48% respectively, but in case (b) it was 78%. Gratifying! However they reported that their toast often went 'boing' rather than 'splat'. Carlo Séquin of the University of California at Berkeley pointed out that the source of the problem is not God's design of the universe, but the 'American Standards Committee for Toast Dimensions', which has clearly decreed that toast be made *the wrong size*. And John Steadman of St. John's College provided a wholly convincing series of arguments to the effect that 'not only is Murphy's

Law a deep consequence of the laws (not just the constants) of nature, but also the laws of nature are a deep consequence of Murphy's Law.' For example the second law of thermodynamics is 'actions have irretrievable consequences,' which is Murphy with a moral dimension, and quantum physics is just a pessimistic version of Murphy: 'if anything can go wrong, *it already has.'*

5

Counting the Cattle of the Sun

'If thou art diligent and wise,
O stranger, compute the number
of cattle of the Sun, who once upon
a time grazed on the fields of the
Thracian isle of Sicily ...' So wrote the
ancient Greek Archimedes to his
friend Eratosthenes of Cyrene.
The answer, first found in 1880, has
206,545 digits. But with some
modern insights into number theory,
and a bit of computer algebra, we
can find an exact formula.

In his 1917 book *Amusements in Mathematics* the English puzzlist Henry Ernest Dudeney describes 'a curious passage in an ancient monkish chronicle' referring to the Battle of Hastings. In 1773 the German dramatist Gotthold Ephraim Lessing published a problem that he had found in the library at Wolfenbüttel: it took the form of 22 elegiac couplets about the Cattle of the Sun and originated with Archimedes around 250 BC. The two puzzles have a common mathematical element, the so-called 'Pell Equation', (mis)named after an obscure seventeenth century English mathematician whose contributions to the area were not original. Archimedes' Cattle Problem has recently been given new life by Ilan Vardi (Occidental College, Los Angeles) with the aid of the computer algebra package Mathematica®. The mathematics involved is itself rather puzzling, and while most of it is classical, some aspects still tax the wits of research mathematicians.

In Dudeney's puzzle 'The men of Harold stood well together, as their wont was, and formed sixty and one squares, with a like number of men in every square thereof ... When Harold threw himself into the fray the Saxons were one mighty square of men, shouting the battle-cries 'Ut!', 'Olicrosse!', 'Godemité!' ' What, asked Dudeney, is the smallest possible number of men there could have been? Mathematically, we want to find a perfect square which, when multiplied by 61 and increased by 1, yields another perfect square (Figure 14). That is, we want integer (whole number) solutions of the equation $y^2 = 61x^2 + 1$. To avoid the trivial answer $x = 0$, $y = 1$ (Harold plus an army of size zero, which is not a 'mighty square of men') we insist that

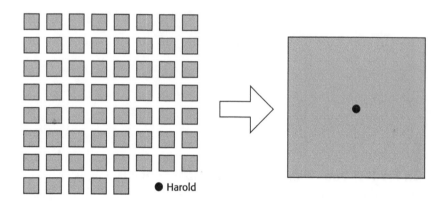

● Harold

Figure 14

If Harold plus his 61 identical squares of yeomen can form a single square, what is the smallest number of men he can have?

x is at least 1. You may wish to try solving this equation before reading on. Don't spend too much time on it, though.

According to the theory developed by mathematicians such as Pierre de Fermat and Leonhard Euler between about 1650 and 1750, equations of this general kind always have infinitely many solutions, where 61 can be replaced by any non-square positive integer. If 61 is replaced by a square, then the equation asks for two consecutive integers to be square, and the only solution is $x = 0$, $y = 1$, which is too trivial to be interesting. The technique for calculating the solutions involves 'continued fractions': it can be found in most number theory textbooks and also in Albert H. Beiler's entertaining *Recreations in the Theory of Numbers* (see Further Reading). As a warm-up, let's take a look at the lesser-known Battle of Brighton (1065), where King Harold's men formed 11 squares, all else being unchanged. Now the equation is $y^2 = 11x^2 + 1$, and a little trial and error reveals the solution $x = 3$, $y = 10$ (Figure 15). That is, $100 = 11 \times 9 + 1$. The next solution is $x = 60$, $y = 199$, and there is a general procedure for finding all solutions once you have found the smallest.

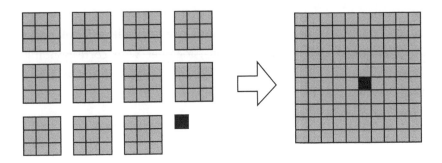

Figure 15

The Battle of Brighton (1065), where King Harold had to solve the equation $y^2 = 11x^2 + 1$.
The solution is $x = 3, y = 10$.

Trial and error will not solve Dudeney's puzzle, though – well, maybe on a computer, but not by hand – because the smallest solution is $x = 226{,}153{,}980$, $y = 1766{,}319{,}049$. Solutions of the Pell Equation $y^2 = Dx^2 + 1$ vary wildly with D. The 'difficult' values of D up to 100 – that is, those that require a value of x larger than 1000 – are $D = 29, 46, 53, 58, 61, 67, 73, 76, 85, 86, 89, 94$, and 97. By far the worst is 61, so Dudeney chose wisely. With a bit of effort, you should be able to find out what happens for $D = 60$ and $D = 62$, on either side of Dudeney's cunning 61: see the answers at the end of this chapter.

Mind you, he could have made the puzzle a lot harder: with $D = 1597$ the smallest solutions are

$x = 13{,}004{,}986{,}088{,}790{,}772{,}250{,}309{,}504{,}643{,}908{,}671{,}520{,}836{,}229{,}100$

$y = 519{,}711{,}527{,}755{,}463{,}096{,}224{,}266{,}385{,}375{,}638{,}449{,}943{,}026{,}746{,}249$

and $D = 9781$ is *much* worse.

The puzzle which Archimedes included in his letter to Eratosthenes begins 'If thou art diligent and wise, O stranger, compute the number of cattle of the Sun, who once upon a time grazed on the fields of the Thracian isle of Sicily ...' In Homer's *Odyssey* there are 350 cattle of the Sun, but Archimedes has a rather larger figure in mind. He states conditions which, in modern notation, boil down to a series of mathematical equations. The herd divides into W white bulls, B black bulls, Y yellow bulls, and D dappled bulls, together with corresponding numbers w, b, y, d of cows. There are seven 'easy' conditions and two much nastier ones. The easy conditions are:

$$W = (1/2 + 1/3)B + Y$$
$$B = (1/4 + 1/5)D + Y$$
$$D = (1/6 + 1/7)W + Y$$
$$w = (1/3 + 1/4)(B + b)$$
$$b = (1/4 + 1/5)(D + d)$$
$$y = (1/6 + 1/7)(W + w)$$
$$d = (1/5 + 1/6)(Y + y)$$

The nastier ones are:

$$W + B = \text{a perfect square}$$
$$Y + D = \text{a triangular number}$$

Here a triangular number is one of the form $1 + 2 + 3 + \ldots + n$, which equals $n(n + 1)/2$.

The first seven equations boil down to a single fact: all the eight unknowns are proportional to each other by fixed ratios. Unravelling the details, we find that all solutions of the first seven equations are of the form

$$W = 10{,}366{,}482n \quad B = 7460{,}514n \quad Y = 4149{,}387n \quad D = 7358{,}060n$$
$$w = 7206{,}360n \quad b = 4893{,}246n \quad y = 5439{,}213n \quad d = 3515{,}820n$$

for any integer n. For details see either Beiler's book, or Vardi's article (Further Reading). Lessing offered his own solution to the puzzle, which amounts to taking $n = 80$, but this does not satisfy the full set of conditions. In 1880 A. Amthor pushed the solution through to a conclusion, and discovered that the total size of the herd is a number with 206,545 digits! He did not calculate this number exactly, but gave the first four digits. Between 1889 and 1893 the Hillsboro, Illinois, Mathematical Club (A. H. Bell, E. Fish, and G. H. Richard) extended this to give the first 32 digits (30 correct). The first complete solution was found by H. C. Williams, R. A. German, and C. R. Zarnke (University of Waterloo) in 1965. The list of all 206,545 digits was first published in 1981, by Harry L. Nelson (Further Reading). He used a CRAY-1 super-computer and the calculation took ten minutes.

Earlier, in 1830, J.F. Wurm had solved a slightly simpler problem, in which the condition that $W + B$ is a perfect square is ignored. (There is an ambiguity in the reading of the original statement of the problem, related to the fact that a bull is longer than it is wide, so bulls can 'form a square' even if their number is not a square. Wurm exploited this loophole.) The condition that $Y + D$ be a triangular number leads, after some algebra, to the requirement that $92,059,576n + 1$ should be a square. The smallest solution of this equation leads to a total number of cattle that is a mere $5916,837,175,686$.

In Wurm's 'solution' the number $W + B$ is not a perfect square. However, there are infinitely many solutions for n, and among them we can seek the smallest that satisfies this missing condition. As Amthor proved, n must be of the form $4456,749m^2$ where m satisfies a Pell Equation:

$$410,286,423,278,424m^2 + 1 = \text{a perfect square.}$$

It now suffices to use the general 'continued fraction' method, whose efficacy was demonstrated by Euler, to find the smallest such m.

There, until recently, the story rested. However, today's mathematics has more sophisticated conceptual tools than those that were available to Amthor, and it also has fast computers that can do arithmetic to hundreds of thousands of digits in the twinkling of an eye. Vardi found that Mathematica can redo all of the above analysis within a few seconds. Pushing a little harder, he found that Mathematica can also produce an exact formula for the size of the herd – something whose existence had not previously been suspected. On a Sun workstation –

an appropriate choice given the owner of the cattle – the computation took an hour and a half. The upshot is that the total number of cattle is the smallest integer that exceeds

$$(p/q)(a + b\sqrt{4729{,}494})^{4658}$$

where

$p = 25{,}194{,}541$

$q = 184{,}119{,}152$

$a = 109{,}931{,}986{,}732{,}829{,}734{,}979{,}866{,}232{,}821{,}433{,}543{,}901{,}088{,}049$

$b = 50{,}549{,}485{,}234{,}315{,}033{,}074{,}477{,}819{,}735{,}540{,}408{,}986{,}340$

There is some debate among scholars about whether Archimedes actually posed the problem at all. The consensus view is that he did, though the text unearthed by Lessing was based on a report by someone else (we don't know who). However, there is little debate about whether Archimedes could have solved his own problem completely. He definitely could not have done: it is too big. Sheer size would not have been an obstacle for Archimedes, whose *Sand Reckoner* includes a number system more than capable of handling a mere 206,545 digits – but calculation by hand would take far too long, even with modern notation.

Did Archimedes have any reason even to know that a solution existed? Probably not. (Indeed even today we do not have a good characterization of those D for which there exist solutions to the Negative Pell Equation $y^2 = Dx^2 - 1$.) Archimedes was certainly clever enough to figure out that some Pell-like equation was required (the Greeks didn't have our algebra, but they had other ways to express such ideas), but he probably could not have been sure that such equations always have solutions. Except that, as David Fowler (University of Warwick) has pointed out (Further Reading), the ancient Greeks also had their own version of continued fractions, so just *possibly* ...

Again, this column resulted in a lot of useful feedback. Chris Rorres of Drexel University told me that more information about the Cattle of the Sun problem (April column) can be found in a paper 'A simple solution to Archimedes' Cattle problem' by A. Nygren of the university of Oulu, Linnanmaa, Oulu, Finland. This describes an algorithm to solve the problem that takes only 5 seconds to run on a Pentium II PC using Maple® or Mathematica®. Links to PDF and postscript files of this paper are on Rorres' web page:

www.mcs.drexel.edu/~crorres/Archimedes/Cattle/Solution2.html

and Maple and Mathematica programs to implement Nygren's algorithm are on:

www.mcs.drexel.edu/~crorres/Archimedes/Cattle/computer2/computer_output.html

The program can be copied to your computer's clipboard, pasted into a Maple or Mathematica worksheet, and run. The smallest solution (nine numbers each of 206,000 decimal digits) can also be found on the same web page.

ANSWERS

$31^2 = 60 \times 4^2 + 1$

$63^2 = 62 \times 8^2 + 1.$

The Great Drain Robbery

When the priceless Robbingham
Rhinoceros was stolen, even the
great Sherlock Holmes was baffled.
Until by chance Dr Watson pointed
out a novel aspect of the geometry
of drains, and that set the great
detective's mind racing. 'When you
have eliminated the impossible,'
he declared, 'then whatever remains,
however improbable, must be the
truth.' But was it?

When I entered Holmes's lodging in Baker Street I found him assembling newspaper, firewood, and coals. It was snowing a blizzard outside, and the room was like an icebox. He rose to his feet and handed me a letter. 'Read this, Watson, and tell me what you think.'

I scanned the page rapidly. 'From the Duke of Robbingham.'

'A rather simple deduction, Watson, since his name is on the letter-head.'

'Excuse me, Holmes, I was merely thinking aloud. He informs you, rather offhandedly I feel, that the Robbingham Rhinoceros has been stolen. "A minor statue, crudely formed, of no great monetary value." My advice, Holmes, is to seek a more challenging case.'

Holmes gave a thin smile. 'Watson, Watson, what can I do to open your eyes? Does the final sentence not strike you as curious, given the admitted unimportance of the stolen statue?'

I reread the letter. It ended: *I request your assistance in locating the stolen property.* 'No, Holmes,' said I. 'It appears entirely normal.'

'The *handwriting*, man!' yelled Holmes. 'Can you not see that the writer was in a state of mortal terror? The descenders on those *p*s are a sure sign, not to mention the loops on the *e*s. I have been of some small service to the Duke of Robbingham in the past, and I am deeply concerned for his safety. Be so good as to hire a special train to Robbingham Hall, while I make preparations.'

During the lengthy journey Holmes amused himself with his violin, while I endeavoured to read a small volume of mathematical conundrums. 'Look, Holmes, here's an interesting one. A man is in the centre of a parallel-sided river 200 yards wide when suddenly a fog descends and he loses all sense of direction. What is the shortest path he can take that will minimize the longest time taken to reach land?'

'He can deduce the direction to the bank by observing the flow of the river,' said Holmes, 'and then swim at right angles to that in a straight line.'

'No, he can't – I mean, suppose it's a lake or someth – '

'But you said it was a river. Oh, very well, what is the path, then?'

'Nobody knows.'

'Wonderful.'

'But they *think* it's one that goes straight for a little over 100 yards, then makes a sharp turn left, followed by a straight bit, a curved arc, and another straight bit (Figure 16a). There's a similar problem when the swimmer is at sea and is 100 yards from a straight shoreline (Figure 16b), and again the answer is only a conjecture.'

'Fascinating,' said Holmes, with sarcasm barely concealed. 'I am glad you are interesting yourself in practical matters, Watson.' He returned to his violin and I tried to continue reading, but my pleasure had been blunted by his disapprobation.

Upon our arrival at Robbingham Hall a maid conveyed us to the Duke's suite. He looked pale and haggard, as if he had suffered a sleepless night. 'Thank you, Lucinda, you may leave us now. Holmes, I am delighted to see you,' he said, in obvious agitation. Holmes endeavoured to calm him, and eventually began to piece the story together. The Robbingham Rhinoceros was a family heirloom, brought from India by the tenth Duke at the end of his successful campaign against the Mad Maharajah of Marzipur. It was made of bronze and worth virtually nothing, but it had a hollow belly with a secret drawer, inside which a number of important documents were traditionally kept. When Holmes enquired of the nature of those documents, the Duke's face went even paler. 'I cannot reveal their contents, Holmes.

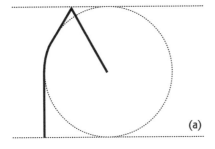

Figure 16

Conjectured best paths to be
followed by a swimmer lost in fog.

(a) When the swimmer is in the
middle of a river with parallel
straight sides.

(b) When the swimmer is a
known distance from a
straight seashore.
(Each dotted line shows one
possible position of a bank or
the shore: the actual position
is some rotation of the line
shown.)

It is an ancient stain on the family escutcheon. If the matter were to be
made public, it would be the end of the Robbinghams.'

'Then we must hope that the animal can be recovered without
further publicity. Show me the room where it was kept.'

The Duke called Lucinda the maid and told her to fetch a lantern. We
set off through the Hall's maze of corridors, to a small, draughty cellar,
decked with cobwebs, its sole light coming from a rusty iron grille in
the ceiling which led to a narrow opening at ground level. There was
an unpleasant smell, and the floor was inches deep in dust. Even I
could see innumerable footprints. In one corner was a large safe. 'The
rhinoceros was in there,' said the Duke.

Holmes studied the floor, his eyes following the trails of footprints.
He took out a magnifying glass and walked across to look closely at the
grille. He inspected the lock on the cellar door and the safe in a similar

manner. Falling to his knees, he searched the dust until he found a small paper wrapper, which seemed to adhere to his fingers. He sniffed the air, glanced towards a pile of old cardboard boxes. 'How large was the statue?'

'Quite big,' said the Duke, holding his hands about three feet apart.

'Then the whole story is here, your grace, for anyone to read who comprehends the runes of observation. At first I feared the rhinoceros might have flown the nest, but now we are faced with a kettle of fish of a different colour.'

The Duke's eyes lit up. I gave Holmes a meaningful look, and he explained his reasoning. 'The cellar door is untouched: the thief made his entrance and exit through the grille. He unlocked the safe and removed the rhinoceros. Lacking knowledge of how to open its secret drawer, and taking into account the difficulty of cutting it open in this cellar, as well as the danger of discovery, he then removed it.'

'But how did he get it out?' asked the Duke. 'It must have been a tight squeeze for the man to get in, and the rhinoceros is considerably larger.'

'Ah. He attached it to an inflatable rubber tube, for buoyancy, and dropped it into the drains, to float away and be collected outside the Hall's grounds.'

'Holmes, that's ridiculous,' I told him. 'You can't possibly know all that. Anyway, there is no drainhole in this cellar.'

'Watson, you underestimate my powers, as usual. Upon the floor I found the remains of a puncture repair kit of the kind used for bicycles. Obviously the tube suffered a puncture while being pulled through the grille, and repairs had to be made on the spot. The smell that you cannot have failed to notice indicates that there is a drain nearby. As for the lack of a drainhole – see for yourself.' Holmes kicked away the cardboard boxes, and a large flagstone appeared with two iron rings in it. 'It had to be there, from the way the footprints led.

'But I have reason to believe the thief was unlucky, Watson. I have made a lifelong study of the smells of drains – you perhaps recall I published a small monograph upon the subject – and I am certain that this one has recently been blocked. Now, Watson, if you will lend me your considerable physical strength, I think we can lift this flagstone.'

In the light of the lantern I saw a deep shaft, lined with stone, about a yard square. At the bottom, a good 40 feet below us, noisome slime lay stagnant.

'The shaft is surprisingly deep, seeing as we are in a cellar,' Holmes muttered.

'The ground rises near the Hall,' said the Duke. 'This cellar is above much of the surrounding land.'

'I see no sign of the rhinoceros,' said I.

'No,' said Holmes. 'But the drains were flowing when the statue was dropped in. Somewhere on the outward journey, the improvised puncture repair gave way and the tube deflated. The rhinoceros then sank to the bottom of the drain, partially blocking it. Other material became caught, and this completed the blockage.'

'So the documents are trapped somewhere in the drains?'

'Precisely. But the shaft is too deep and dangerous for anyone to attempt to locate the blockage from this end. We must break into the drainage system at a more convenient point. Do you have maps?'

'In the library,' replied the Duke. But none of the maps showed a drain that might plausibly connect to the cellar. 'I swear there *was* such a map,' said the Duke, puzzled.

'It must be missing,' Holmes deduced.

'Drat,' said I. 'The scoundrel may even now be making his way back up the drains from the exit, seeking his booty.' A thought struck me. 'Holmes, perhaps he has already done so!'

'No, for then the drain would have become at least partially unblocked again,' said Holmes. 'The thief must also have had trouble finding an alternative entrance. But he might very well make the attempt tonight, so there is no time to lose.' He paused, deep in thought. 'When we arrived, I saw an elderly gentleman hoeing the carrot-beds.'

'Old Ned, you mean. Deaf as a post, but a good servant. Been with us for ages.'

'Perhaps he can recall the layout of the drains. Gardeners often remember that sort of thing.'

After much gesticulating and shouting, Holmes managed to explain to Old Ned what was wanted. 'Ar, yur,' said Ned. 'I heard tell there be a big old drain running roight across the front lawn in a dead straight line. But nobody knew where it went – though the previous kitchen-maid but six did once tell me that upstream it zigzagged under the cellars. Downstream, she said, it carried on straight as an arrow.'

'Can you show us where this drain runs?' asked Holmes.

'Ar, no. But I do recall it passed within a hundred yards or less of the statue of the water-nymph.'

'We must dig a trench,' said the Duke. I will summon every available man.'

'We must also dig it *fast*,' said Holmes.

'And the right shape,' said I. 'Otherwise it might not encounter the drain at all.'

'What we need to know,' said Holmes, 'is the shortest trench that is guaranteed to meet *every* straight line that passes within a hundred yards of the statue of the water-nymph.' See Figure 17.

'We could dig a circular trench with a hundred-yard radius,' the Duke suggested.

'With a length of 200π yards, or about 628 yards,' Holmes rapidly calculated.

'I doubt we have time for such a long trench,' said the Duke. 'My men could get close to that figure, though. Can we do better?'

'How about a straight line cutting across the Duke's circle, of length 200 yards?' I asked.

'Excellent, Watson,' said Holmes, 'except that such a trench misses many possible positions for the drain.'

'Two such lines at right angles? Length 400 yards?'

'Same problem, Watson. No, we must think this through more care-

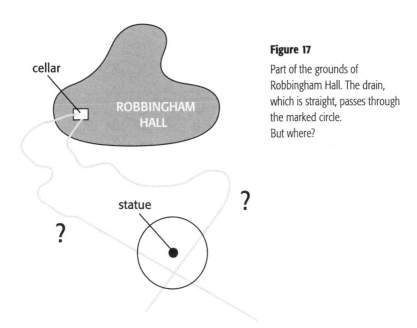

Figure 17

Part of the grounds of Robbingham Hall. The drain, which is straight, passes through the marked circle. But where?

fully. Mathematically, we are looking for the shortest curve that meets every chord of a circle of radius 100 yards – a chord being any straight line that meets the circle. We must of course include the tangents to the circle, which meet it at only one point.'

'Why, Holmes?'

'Because old Ned said 'a hundred yards or less', implying that a distance of 100 yards might have occurred, and that is the distance to a tangent.'

'Ah,' said I, impressed by his incisive logic. 'But that is a very complicated problem, for there are many chords to consider. Might we not simplify it somehow?'

Holmes's head jerked back in surprise. 'An excellent suggestion, Watson, your best for weeks. It is of course sufficient to consider only the tangents. A curve that meets all tangents necessarily meets all chords.'

'Why?'

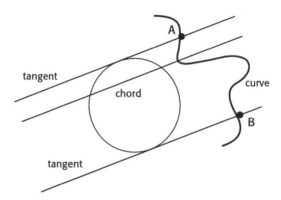

Figure 18

If a curve meets every
tangent, then it also
meets every chord.

'Choose any chord, and consider the two tangents parallel to it
(Figure 18). The curve meets each of these, say at points A and B. By
continuity the part of the curve that joins A to B must cut the chord.' He
rubbed his chin in deep contemplation. 'I *almost* have it, Watson. But
there is a gap in my reasoning.'

'Out with it, man!'

'Very well. There is a class of curves that automatically meet every
tangent, and I can find the shortest of those. They are curves that start
and end on opposite sides of the *same* tangent to the circle, wrapping
all the way round the circle in between, and remaining outside or on
the circle. Let me call such a curve a *strap*, since it straps that particular
tangent to the circle.' See Figure 19a.

'So we need to find the shortest strap?'

'That part is simple. First, observe that the strap must meet the circle
somewhere, or else it could be 'tightened down' towards the circle,
thereby becoming shortened. Suppose that it first meets the circle at a
point B, and last meets it at C. Then AB and CD must be straight lines,
or else the strap can be shortened along those segments. Moreover, BC
is just a single arc of the circle, for similar reasons.' See Figure 19b.

'I think that the lines AB and CD must be tangent to the circle,' I put
in. 'Otherwise the curve could be shortened by moving B and D to
positions where they are indeed tangents.' See Figure 19c.

'Of course, Watson. Finally, we must ask where the points A and B

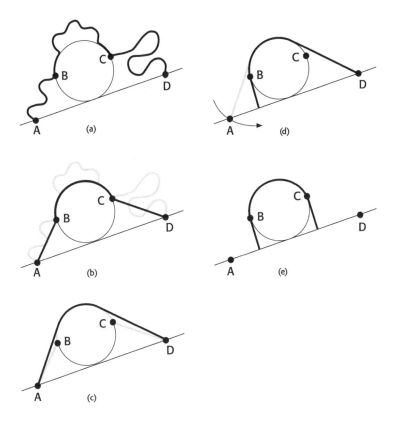

Figure 19

(a) A curve that wraps round the circle starting and ending on the same tangent always meets all tangents to the circle. Call such a curve a strap.

(b) Lines AB and CD can be straightened, and excursions from the circle between B and C eliminated.

(c) The strap gets shorter when AB and CD are made tangent to the circle.

(d) If DAB is not a right angle, we can swing end A round to shorten the strap. Similarly for end D.

(e) The shortest strap consists of a semicircle and two tangential straight lines of length equal to the radius. In fact, this is the shortest curve that meets every tangent to the circle (hence also every chord).

should be located. It seems to me that both AB and DC must be at right angles to the tangent AD. For if not, then we can 'swing' the strap into a right angle position, and again it will become shorter.' See Figure 19d.

'Yes, yes,' I cried. 'And then the arc BC is a semicircle, and we have found the shortest curve.' See Figure 19e.

'Unfortunately we have found only the shortest *strap*,' said Holmes. 'Though I find it hard to see how any other curve could be shorter.'

We stood in silence for a few minutes, until an idea struck me. 'All is not lost,' I cried. 'How long is this shortest strap?'

'$(2 + \pi) \times 100$, or about 514 yards in this case,' said Holmes.

'A saving of 89 yards,' I said meaningfully. 'There is no time to lose: I will ask the Duke to set his men to work.'

While they dug, Holmes and I continued to seek even shorter curves, but we could find none. Then I had a sudden recollection. 'The book, Holmes! The book I was reading on the train!' I pulled it from my coat pocket. 'Let me see ... yes, the shortest path that meets every chord of a circle does indeed consist of two parallel straight segments and a semicircle.'

Holmes took a look at my book. 'I confess I misjudged the utility of this slim volume, Watson,' he said. 'But, out of interest, how do we *know* that the path proposed is truly the shortest?'

'Why, it has been proved beyond any shadow of doubt,' I told him. 'By several different people, under slightly differing conditions on the nature of the curve allowed, but first by H. Joris. However, all known proofs are surprisingly long and complex. It would be of great interest if anyone could find a short, simple proof.'

We took to discussing other variations on such problems – a man lost in a rectangular forest of known dimensions, a skater on a frozen elliptical lake when a blizzard comes down ... Suddenly Old Ned let out a yell; he had located the drain. A few minutes' work made its direction clear, and Holmes sighted along it. 'Ah, yes, beyond that distant coppice, where the stream flows. That is where we shall find the outlet. And the thief, when he comes to secure his booty.'

HOW TO FIND THE SHORTEST TRENCH

A complete proof of Watson's claim that the shortest trench consists of a semicircle and two straight segments is surprisingly complicated (see Further Reading). However, it is relatively straightforward to make it plausible. First, note that if a curve meets every tangent to a circle then it must also meet every chord. To see why, consider the two tangents parallel to the chord (Figure 20). The curve meets each of these, say at points A and B. By continuity the part of the curve that joins A to B must cut the chord.

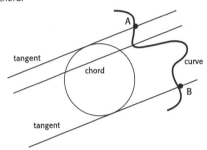

Figure 20

If C meets every tangent, then it also meets every chord.

Thus it is enough to find the shortest curve that meets every *tangent* to the circle. This simplifies the proof because we can concentrate on the tangents, which are simpler to deal with, and forget all the other chords.

Since no tangent meets the interior of the circle it is plausible that the shortest curve remains either outside or on the circle. This is true, but not entirely obvious. Suppose the curve cuts across the circle. Then that part of the curve that lies within the circle is 'wasted', but we cannot simply remove it because the curve then falls into two parts. We have to show that the curve can be re-routed outside the circle, without increasing its length, so that it still cuts every tangent. This requires a complicated case-by-case analysis and is the hardest part of the proof.

Having established this, we then show that the curve must start and end on opposite sides of the *same* tangent to the circle, wrapping all the way round the circle in between, as in Figure 21. (The curve may run along the circle in some places.) Moreover, any curve

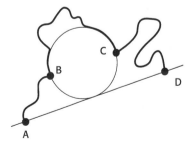

Figure 21

General form for the shortest curve.

of that form automatically meets every tangent. So all we need do is find the shortest curve of that form.

Having pinned the geometry down to this very specific set of curves, it is now easy to show that AB and CD must be straight lines, and BC is just a single arc of the circle. It then follows that AB and DC are at right angles to the tangent AD, and the arc BC is a semicircle. (Intuitively, imagine the curve ABCD to be made of elastic, with A and D free to slide along the tangent AD. As the elastic contracts, ABCD automatically takes up the shortest curve. Clearly the part between B and C will wrap tightly round the circle, while AB and CD will contract to straight lines. That DAB and ADC are right angles is almost as obvious.)

⬡⬡

'You think the thief is a man, then, Holmes?'

'The footprints make that plain, Watson.'

We secreted ourselves among the trees, and settled down to wait. No sooner had the Sun set than we heard footsteps, and the sounds of splashing. A masked figure came into view, and Holmes leaped on him. There was a brief struggle, and then the criminal was lying on his back and Holmes was sitting on top. He seized the mask, and pulled.

'My god, it's Lucinda,' said the Duke, who had just arrived on the scene. 'What are you doing here, my dear? Don't tell me *you* are our thief?'

'God please your grace, no. Last Monday evening I was sent to the cellar to sort out some boxes of junk, but the door was locked and I couldn't find the key, Ned had been mending his bicycle in the cellar, and I think he forgot to hand the key back when he'd finished. Anyway, I climbed in through the grille. There was an open drain hole with a stone slab beside it. The safe was open and I saw this funny old rhinoceros statue thing in it, and I pulled it out to take a look. It was heavier than I expected, and it accidentally fell down the drain. A lot of junk went in with it, and the drain got blocked. I looked down and I couldn't see the statue, so I assumed it had floated away before the blockage occurred.'

'Lucinda, it's made of *bronze*.'

'I thought it was wood with gold paint, sir. Anyway, I panicked. I pulled the stone over the drain hole, covered it with some old boxes, shut the safe, and climbed out again. This morning I found a map in the library showing where the drains went – it's hidden in my room, I swear I always intended to give it back once I'd – anyway, I was about to crawl up the drain looking for the statue when this gennelmun – ' she gave Holmes a winsome smile – 'leaped on me.'

'So the Robbingham Rhinoceros has never moved from the bottom of the shaft in the cellar,' the Duke mused, 'and I have a five-hundred yard trench in a lawn that has been carefully tended for six centuries.' He gave Holmes a hard stare.

'It is a question of logical deduction,' said Holmes. 'As I have often stated, when you have eliminated the impossible, then whatever remains, however improbable – '

'Yes, Holmes, yes! Go on!' I cried. 'That's brilliant!'

' – remains improbable,' said Holmes, deflated. 'There's probably something altogether different going on, and you've missed it. But don't quote me on that,' he warned.

'My lips are sealed,' I replied. But my pen and notebook are not. After all, a biographer has to make a living.

Two-Way Jigsaw Puzzles

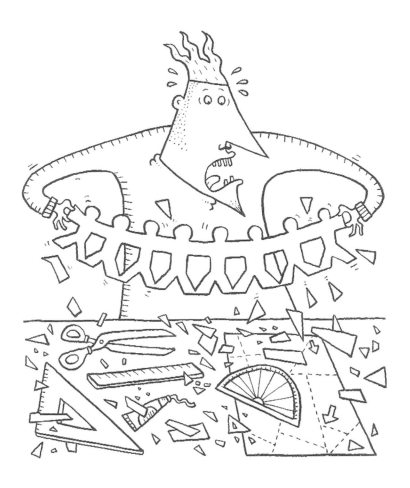

A rivalry between the great puzzle-masters of Britain and America sets the scene for a classic geometric recreation: dissections. Can you cut up a regular hexagon and rearrange the pieces to form a regular octagon? Or turn a pentacle into a square? A beautiful book by Greg Frederickson, an expert dissector, has revealed some of the tricks of the trade.

Early in their careers, the great puzzlists Sam Loyd and Henry Ernest Dudeney – one American, one English – collaborated to write a regular puzzle column for the magazine *Tit-bits*. Loyd wrote the puzzles, and Dudeney (under the pseudonym 'Sphinx') provided the commentary and awarded prizes. Collaboration soon turned into rivalry, and the two men went their separate ways. In so doing, they created an entire puzzle industry on both sides of the Atlantic, by formulating tantalizing mathematical questions within simple but engaging stories.

A typical example of their work is Loyd's Sedan Chair Puzzle (Figure 22a). The mathematical problem is to cut the sedan shape into as few pieces as possible, and reassemble them to form a square; Loyd embeds it in a tale in which the young lady's sedan chair folds up, cunningly, to protect its occupant from the rain. The answer (Figure 22b) is neat, elegant, and far from obvious.

Puzzles of this kind are known as 'dissections'. I prefer to think of them as jigsaw puzzles with more than one answer. A wonderfully entertaining book on this time-honoured mathematical recreation is Greg N. Frederickson's *Dissections: Plane and Fancy* (Further Reading). Every puzzle enthusiast and amateur mathematician should own a copy. To whet your appetite, Figure 23 shows a selection of interesting examples, all from Frederickson's book.

The basic mathematical concept that underlies all dissection puzzles is area. When a shape is cut up and the pieces are rearranged, the total area does not change. Some very deep mathematics indeed lies behind this simple and apparently obvious statement. In particular it is *false*

THE SEDAN CHAIR PUZZLE.

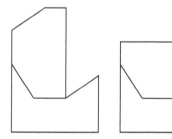

Figure 22

(a) Sam Loyd's Sedan Chair puzzle.

(b) Its solution.

in three dimensions if the nature of the 'pieces' is allowed to be sufficiently complicated. In the celebrated Banach–Tarski Paradox a solid sphere is 'dissected' into six pieces, which can be reassembled to form two solid spheres, each the same size as the original. Stefan Banach and Alfred Tarski collaborated on this weird theorem in 1924. It isn't really a paradox: it's a perfectly sensible, logically valid result. But it seems so bizarre that 'paradox' has stuck.

I hesitate to mention this strange result, because it is such an obvious impossibility: how can the volume *double*, just by rearranging the pieces? The trick is to employ pieces that are so strange that they do not possess a well defined volume. Indeed it is stretching the language to

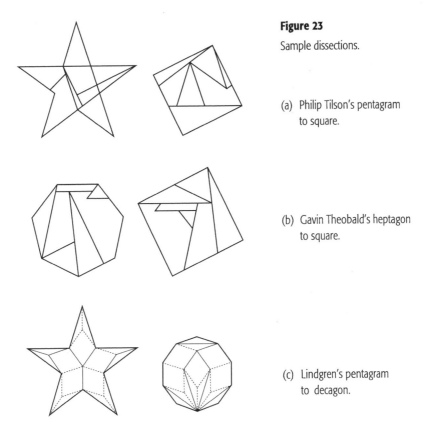

Figure 23

Sample dissections.

(a) Philip Tilson's pentagram
to square.

(b) Gavin Theobald's heptagon
to square.

(c) Lindgren's pentagram
to decagon.

describe them as 'pieces', for they are not single connected objects – they are more like infinitely complex spherical dust clouds. The ideas are summarized in my book *From Here to Infinity* and described in all their glory in Stan Wagon's *The Banach–Tarski Paradox* (Further Reading).

There is of course no practical way to realize this theoretical dissection with a physical object – so traders in precious metals can breathe a sigh of relief – but it does demonstrate how subtle the concept of volume is. Even to describe the mathematical construction requires many pages of argument and several sophisticated ideas. Curiously, no 'paradoxical' dissections, in which areas change, are possible in plane geometry, no matter how complex the pieces may be – as Tarski proved in 1925. However, they *are* possible on the surface of a sphere.

When the pieces into which the object is cut are nice enough to have well-defined areas or volumes – and especially when they are polygons or polyhedrons, with straight edges and faces – there are no intuition-bending constructions like that of Banach and Tarski. Indeed in 1833 P. Gerwein, a lieutenant in the Prussian army, answered a basic question about dissections raised by the Hungarian mathematician Wolfgang Bolyai. Gerwein proved that given any two plane polygons of equal area, there is a finite set of identical polygonal pieces that can be assembled to form either shape. This result is traditionally called the Bolyai–Gerwein Theorem – although in fact it seems to have first been proved by William Wallace in 1807.

The Bolyai–Gerwein Theorem does *not* generalize to three dimensions. In 1900 David Hilbert asked whether any two polyhedrons of equal volume are 'equivalent by dissection' – can be assembled from the same set of polyhedral pieces. One year later Max Dehn proved the startling result that a cube and a regular tetrahedron of equal volume are *not* equivalent by dissection.

The real fun, of course, comes in finding neat or surprising examples of shapes that *are* equivalent by dissection. You can make some progress by inspired trial and error, but only if you have a vivid spatial imagination. One of the great virtues of *Dissections: Plane and Fancy* is that, as well as exhibiting a vast number of dissections, it explains many of the general principles involved in finding them.

One is the 'step principle' (Figure 24), where a shape is cut along a 'staircase' which can then be moved one step along to create a different

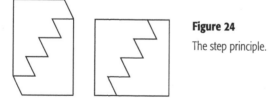

Figure 24
The step principle.

shape. Both Loyd and Dudeney made use of this principle in many of their puzzles. David Collison, a dissection enthusiast who was born in England and worked as a computer programmer and consultant in the United States, devised elaborate dissections based on this principle. Figure 25 shows one of his creations, a dissection proof of the well known 'Pythagorean' fact that $5^2 + 12^2 = 13^2$, realized in the context of pentagons. You can see several distinct uses of the step principle in this example.

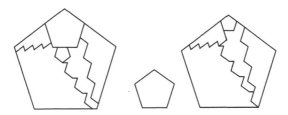

Figure 25 Collison's Pythagorean dissection. Here the largest pentagon has area 169 square units; the smaller ones have areas 25 and 144 respectively.

Another general method is the 'tessellation principle'. Many interesting shapes can be embedded in tessellations – tiling patterns that cover the plane. If two different tessellations, each formed from tiles of the same area, are superposed, it often becomes possible to 'read off' a dissection from one shape to the other. Figure 26 shows a simple example, in which one shape is a Greek cross and the other is a square.

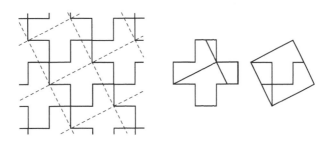

Figure 26 Tessellation: Greek cross to square.

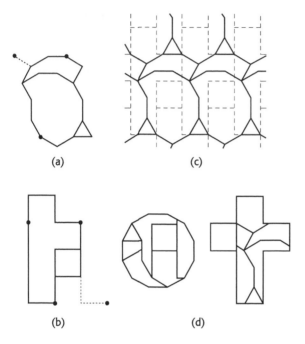

Figure 27

Lindgren's dodecagon to Latin cross.

(a) Cut the dodecagon into three pieces and rearrange to form a tile.

(b) Cut up the Latin cross.

(c) Tile the plane two ways.

(d) Comparing the two tilings leads to the dissection of the dodecagon into a Latin cross.

Figure 27 shows a more elaborate use of the same basic idea. This dissection of a dodecagon to a Latin cross is due to Harry Lindgren, a world expert in dissections and author of *Geometric Dissections* (Further Reading). The first step – and the hardest – is to cut the dodecagon into three pieces that can be rearranged to form a rather complicated tile (Figure 27a). The Latin cross of equal area to the dodecagon can also be cut into two pieces (Figure 27b). Each of these shapes tiles the plane (Figure 27c). Comparing the two tilings leads to the dissection of Figure 27d.

Yet another variation on the tessellation principle lies behind a dissection published by George Biddle Airy, who served as the British Astronomer Royal from 1836 to 1881, to prove Pythagoras's Theorem (Figure 28). The little poem is Airy's.

A third general method is the strip principle. The two shapes are cut into pieces that between them can tile an infinitely long strip. If the

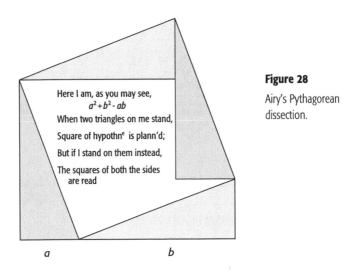

Here I am, as you may see,
$a^2 + b^2 - ab$
When two triangles on me stand,

Square of hypothn^e is plann'd;

But if I stand on them instead,

The squares of both the sides
are read

a b

Figure 28

Airy's Pythagorean
dissection.

strips are then overlapped, they determine a dissection. Figure 29 shows how Paul Busschop's dissection of a hexagon to a square can be derived by the strip method. Busschop was a Belgian who wrote a book on peg solitaire, published posthumously in 1879 by his brother.

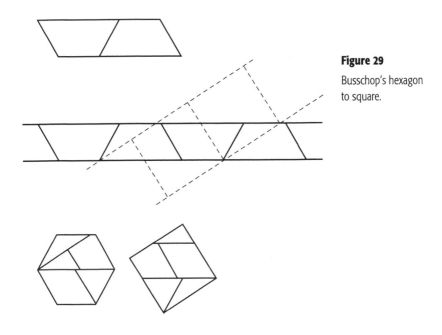

Figure 29

Busschop's hexagon
to square.

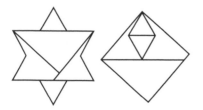

Figure 30
Bradley's star of David
to square.

Figure 30, dissecting a star of David to a square, can be obtained by the same method. It was invented by Harry Bradley, an American engineer who was an instructor at MIT in 1897.

Dissection puzzles are not confined to changing just one shape to another. Often an entire set of shapes must be cut up and reassembled. Figure 31 shows two examples: four octagons to one, and six dodecagons to one.

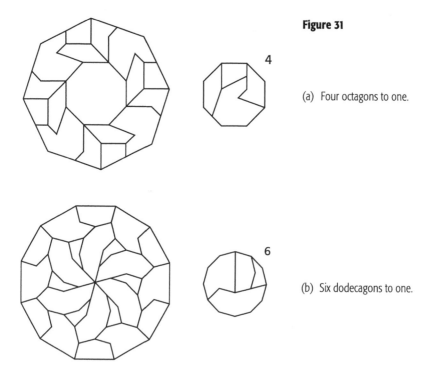

Figure 31

(a) Four octagons to one.

(b) Six dodecagons to one.

Dissections are so compelling that occasionally the eye can mislead the head. In 1901 Loyd made a memorable error – described by Frederickson as 'perhaps his biggest goof' – when he claimed to dissect a mitre (a square with one quarter removed) into a square as in Figure 32a. Unfortunately, the apparent 'square' is actually a rectangle whose sides are in the proportion 49:48. Ironically, Loyd called this the Smart Alec Puzzle. His rival Dudeney pointed out the error in 1911, and gave a correct dissection, Figure 32b. So if you want to look for your own dissections, take the advice given by many a parent to their offspring: 'Have fun – but be careful.'

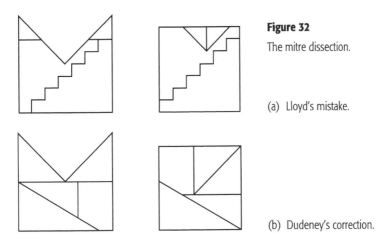

Figure 32
The mitre dissection.

(a) Lloyd's mistake.

(b) Dudeney's correction.

8

Tales of a Neglected Number

The 'golden number' 1.618034
is a staple of recreational
mathematics, turning up in
pineapples, spirals, and an
800-year old problem about
exponential growth of a population
of rabbits. Here we take a look at
a lesser-known, but equally
interesting relative (of the number,
not the rabbits), namely 1.324718,
which the architect Richard Padovan
calls the 'plastic number'.

A

lan St. George is a mathematical sculptor, who often makes use of the well-known 'golden number'. In 1995–6 he exhibited a number of his works, and the catalogue of his exhibition mentions a less celebrated relative, referring to a series of articles in which 'The architect Richard Padovan revealed the glories of the "plastic number" ... The plastic number has little history, which is strange considering its great virtues as a design tool, but its provenance in mathematics is almost as respectable as that of its golden cousin ... It doesn't seem to occur so much in nature, but then, no one's been looking for it.'

I found this intriguing and decided to dig out more about this curious, little-known number. I suspect it has been rediscovered many times, and probably has a variety of names.

For purposes of comparison, let me start with the golden number ϕ. (An excellent source is Mario Livio's book in Further Reading.) This satisfies the equation $\phi = 1 + 1/\phi$, which implies that $\phi = (1 + \sqrt{5})/2 = 1.618034$ approximately. The golden number is intimately associated with the geometry of pentagons – it is the ratio of the diagonal of a pentagon to its side – and hence is also associated with the dodecahedron and icosahedron. It has close connections with the celebrated Fibonacci numbers. To see this connection look at Figure 33, a spiralling system of squares. The initial square (marked in black) has side 1, as does its immediate neighbour. Then a square of side 2 is added, to fit snugly against the first two, followed in turn by squares of side 3, 5, 8, 13, 21, and so on. These are the Fibonacci numbers, each being the sum of the

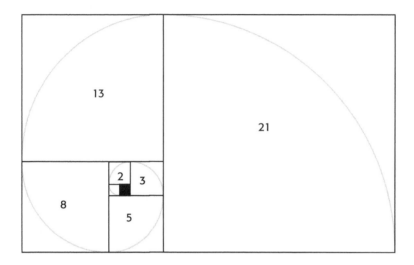

Figure 33 Spiral of squares forms Fibonacci numbers.

previous two. This is clear from the figure – for example the square of side 21 has the same height as those of sides 13 and 8 put together.

The ratio of consecutive Fibonacci numbers tends to the golden number. For example $21/13 = 1.615384$. This fact is a consequence of the rule for generating Fibonacci numbers, in combination with the equation $\phi = 1 + 1/\phi$. In Figure 33 I have added a quarter circle inside each component square, fitting together to form an elegant spiral. This spiral is a good approximation to the so-called 'logarithmic spiral' which is often found in nature, for example in the shell of a nautilus. Successive turns of the spiral grow at a rate approximately equal to the golden number.

That's the golden tale: now for the analogous plastic one. We start with a diagram much like Figure 33, but composed of equilateral triangles (Figure 34). The initial triangle is marked in black and successive triangles spiral in a clockwise direction: the spiral shown is again roughly logarithmic. In order to make the shapes fit, the first three triangles all have side 1. The next two have side 2, and then the

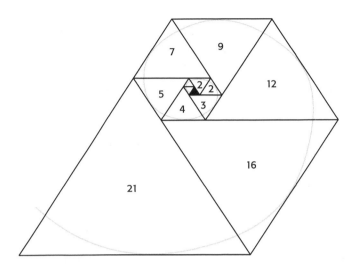

Figure 34 Spiral of triangles forms Padovan numbers.

numbers go 4, 5, 7, 9, 12, 16, 21 and so on. Again there is a simple rule of formation, analogous to that for Fibonacci numbers: now each number in the sequence is the sum of the previous number *but one*, together with the one before that. For example 12 = 7 + 5, 16 = 9 + 7, 21 = 12 + 9. Again this is clear from the geometry of the triangles and the conditions needed for them to fit together. Let me call this sequence the *Padovan sequence* in honour of Richard Padovan.

Actually, he asked me not to do this because he didn't invent the number, but then, Pell didn't invent (or solve) Pell's Equation either, and Bolyai and Gerwein were not the first to prove the Bolyai–Gerwein Theorem. You'll see that I have a good reason – an Italian joke. And people deserve credit for making ideas *interesting*. So, with profuse and heartfelt apologies to Richard, I will stick to my historically inaccurate terminology. Needs must when Journalism drives.

It is curious that 'Pádova' is the Italian form of 'Padua', and Fibonacci was an Italian – from Pisa, roughly a hundred miles away from Padua. 'Fibonacci' allegedly means 'son of Bonaccio'; though the name he himself used was 'Leonardo of Pisa' and his famous nickname seems not to have been invented until the nineteenth century. I am sorely tempted to rename the Fibonacci numbers the 'Pisan sequence' to preserve the flavour of Italian geography, but I have just about managed to resist the temptation and retain the traditional nomenclature.

The first 20 terms of the Fibonacci sequence $F(n)$ and Padovan sequence $P(n)$ are listed in Table 1. In algebraic form the generating rules are

$$F(n + 1) = F(n) + F(n - 1) \qquad \text{where } F(0) = F(1) = 1$$
$$P(n + 1) = P(n - 1) + P(n - 2) \qquad \text{where } P(0) = P(1) = P(2) = 1$$

making the family resemblance very visible. The table also includes the 'Perrin numbers' $A(n)$, to be discussed a little later.

Table 1

n	F(n)	P(n)	A(n)	n	F(n)	P(n)	A(n)
0	1	1	3				
1	1	1	0	11	144	16	22
2	2	1	2	12	199	21	29
3	3	2	3	13	343	28	39
4	5	2	2	14	542	37	51
5	8	3	5	15	885	49	68
6	13	4	5	16	1427	65	90
7	21	5	7	17	2312	86	119
8	34	7	10	18	3739	114	158
9	55	9	12	19	6051	151	209
10	89	12	17	20	9790	200	277

If you take ratios of successive terms of the Fibonacci sequence, such as 8/5, 13/8, 21/13, and so on, you will find that the ratios approach the golden number as their limit. For instance, $21/13 = 1.6153$, $34/21 = 1.6190$, and $9790/6051 = 1.6179$. There are simple ways to prove this. For similar reasons, the plastic number, which from now on I shall call p, and whose approximate value is 1.324718, arises as the limit of the ratio of successive Padovan numbers. Thus $200/151 = 1.3245$. The rule of formation for the Padovan numbers leads to the equation $p = 1/p + 1/p^2$, or equivalently $p^3 - p - 1 = 0$. The number p is the unique real solution of this cubic equation.

The Padovan sequence increases much more slowly than the Fibonacci sequence because p is smaller than ϕ. There are many interesting patterns in the Padovan sequence. For example, Figure 34 shows that $21 = 16 + 5$, because triangles adjacent along a suitable edge have to fit together; similarly $16 = 12 + 4$, $12 = 9 + 3$, and so on. This tells us that $P(n + 1) = P(n) + P(n - 4)$, an alternative rule for deriving further terms of the sequence.

Correspondingly the equation $p = 1 + 1/p^4$, or $p^5 - p^4 - 1 = 0$, must be satisfied – and it is not immediately obvious by algebra that p, defined as a solution of a cubic equation, must also satisfy this *quintic* (fifth degree) equation. You might like to puzzle out why this is.

In the original column I pointed out that some numbers, such as 3, 5, and 21 are both Fibonacci and Padovan, and asked: are there others? If so, how many: is that number finite or infinite? Benjamin de Weger (Further Reading) proved that these are the *only* numbers that are both Fibonacci and Padovan, together with the trivial cases 0, 1, and 2. I also noted that some Padovan numbers, such as 9, 16, and 49 are perfect squares, asking: are there others? How many? The square roots here are 3, 4, and 7 – also Padovan numbers. Is this a coincidence or a general rule? These questions remain open, and deserve further study.

Another way to generate the Padovan numbers is to mimic the use of squares for Fibonacci numbers, but using cuboids – three-dimensional boxes with rectangular faces. Now we get a kind of three-dimensional spiral of cuboids. In more detail, start with a cuboid of side 1, and place another adjacent to it. The result is a $1 \times 1 \times 2$

cuboid, and on the 1 × 2 face we add another 1 × 1 × 2, getting a 1 × 2 × 2 cuboid. On to a 2 × 2 face add a 2 × 2 × 2 cube, to form a 2 × 2 × 3 cuboid overall. On to a 2 × 3 face add a 2 × 2 × 3 to get a 2 × 3 × 4 overall, and so on. Continue the process, always adding cuboids in the sequence east, south, down, west, north, up. At each stage the new cuboid formed will, overall, have as its sides three consecutive Padovan numbers.

A sequence with the same rule of formation as the Padovan numbers, but using different starting values, was studied by the French mathematician Édouard Lucas in 1876. In 1899 his ideas were developed by R. Perrin, and the sequence is now known as the *Perrin sequence A(n)*. The Perrin numbers differ from the Padovan numbers in that $A(0) = 3$, $A(1) = 0$, $A(2) = 2$: their values are also shown in the table. Again the ratio of consecutive Perrin numbers tends to p, but Lucas noticed a more subtle property. Whenever n is a prime number (one having no factors other than itself and 1), n divides $A(n)$ exactly. For example 13 is prime, $A(13) = 39$, and $39/13 = 3$. Similarly 19 is prime, $A(19) = 209$, and $209/19 = 11$.

This theorem provides a curious test for a number to be composite, that is, not prime. For example when $n = 18$ we have $A(18) = 158$, and $158/18 = 8.777$, which is not a whole number. Therefore 18 must be composite. So we can use Perrin numbers to test for non-primality: any number n that does not divide $A(n)$ is composite. One curious feature of this test is that it does not actually exhibit a divisor of n, which is the most obvious way to prove that a number is not prime. Of course here we know that $18 = 2 \times 3 \times 3$, but for bigger numbers the factors may not be so obvious: nevertheless you can still divide $A(n)$ by n and see what you get.

If n divides $A(n)$, must n always be prime? This does *not* follow from Lucas's theorem – any more than 'if it rains then I get wet' implies 'if I get wet then it rains'. (I might have fallen into a pond on a perfectly dry day.) And in fact, the answer is 'no'. In 1982 Jeffrey Shallit (University of Waterloo) found two composite numbers that divide the corresponding Perrin number, namely 271,441 and 904,631. Later, he found a third, 16,532,714. A more complicated test for primality, based on several successive Perrin numbers, has also been studied, and no exceptions are currently known.

9

Is Monopoly Fair?

No rent coming in, in hock to the
bank, with your entire financial future
hanging on a throw of the dice ...
Sounds just like the stock market
when the dotcoms crashed.
But of course it's Monopoly®.
A game, without doubt – but is it a
mathematical game? Indeed it is, as
this first leg of a double-header aims
to prove. In this warm-up, I'm not
going to tackle the real game –
just a simplified version. See the
next chapter for the gory details.

There can scarcely be a family that has not played Monopoly®, probably the world's most famous board game, involving luck, strategy, and cut-throat economics. There is some very interesting and quite deep mathematics associated with all such board games: they form what a probability theorist would call a Markov chain, a concept named after the Russian mathematician Andrey Andreyevich Markov who invented a general theory of such things in the early 1900s.

I'm not going to remind you about all the rules of Monopoly, but we need to know that players take their turn to throw a pair of dice, and the total number of spots determines how many squares they move. There is a rule that a player throwing a double throws again (but if they throw three doubles in a row they go to jail), and one simplification that I shall make here is to ignore this rule. The same kind of analysis can still be applied if you don't, but the mathematics gets much more complicated – and it's complicated enough already.

Players start from the square known as 'Go'. Anyone who has played games involving two dice knows that some totals are more likely than others. In fact the most likely total is 7, with probability 1/6. This happens because there are six ways to total 7, namely 1 + 6, 2 + 5, 3 + 4, 4 + 3, 5 + 2, 6 + 1, out of the 36 possible combinations of the two dice, so the probability of a 7 is 6/36 = 1/6. Next come 6 and 8 (probability 5/36), then 5 and 9 (probability 4/36 = 1/9), 4 and 10 (probability 3/36 = 1/12), 3 and 11 (probability 2/36 = 1/18), and finally 2 and 12 (probability 1/36). So the first player is most likely (over many games) to land on the seventh square along, which is a

Chance square – meaning that they must draw, and obey, a card from the Chance pack. Again, to keep the analysis simple, I'll start by assuming that the instructions on the card are *not* obeyed. On either side of that square, only slightly less probable, are The Angel, Islington (Oriental Avenue in the US) and Euston Road (Vermont Avenue). So the first player has an excellent chance of securing one of these two desirable properties, and if so, the other players' chances of buying a property on their first throw are reduced.

This is no doubt one of the reasons why the game's designers put the cheap properties (with low rents) near the start. The really expensive, but lucrative, Park Lane (Park Place) and Mayfair (Boardwalk) are several turns round the board, by which time – presumably – the probabilities have evened up.

But have they?

There are two versions of this question. One is a simplified version which does not apply to the real game without significant changes, but has the virtue of illustrating the Markov chain approach in a case where the equations can easily be solved. The second is the real question for the real game, with its numerous complications; this also yields to the Markov chain approach, but the calculations are extensive and require a computer. I'll tackle the simplified version in this chapter, to explain the principles, and move on to the real thing in the next chapter.

Oh, and to avoid more parenthetic translations, here's a table showing how the names differ on either side of the Atlantic. (There are probably hundreds of national variants of the board, plus some media-tie-in ones too – no room to list them all.)

In order to tackle the simplified version of my question, I shall introduce yet a further simplification. Instead of considering a throw of two dice, I will imagine them being thrown one at a time. From this point of view each player makes two moves: an initial 'ghost' move – on which they ignore what is written on the resulting square – followed by a second 'real' move. Because our interest is the flow of

Table 2

British	American	British	American
GO	GO	FREE PARKING	FREE PARKING
OLD KENT ROAD	MEDITERRANEAN AVE	STRAND	KENTUCKY AVE
COMMUNITY CHEST	COMMUNITY CHEST	CHANCE	CHANCE
WHITECHAPEL ROAD	BALTIC AVE	FLEET STREET	INDIANA AVE
INCOME TAX	INCOME TAX	TRAFALGAR SQUARE	ILLINOIS AVE
KINGS CROSS STN	READING RAILROAD	FENCHURCH ST STN	B&O RAILROAD
THE ANGEL ISLINGTON	ORIENTAL AVE	LEICESTER SQUARE	ATLANTIC AVE
CHANCE	CHANCE	COVENTRY STREET	VENTNOR AVE
EUSTON ROAD	VERMONT AVE	WATER WORKS	WATER WORKS
PENTONVILLE ROAD	CONNECTICUT AVE	PICCADILLY	MARVIN GARDENS
JUST VISITING/JAIL	JUST VISITING/IN JAIL	GO TO JAIL	GO TO JAIL
PALL MALL	ST CHARLES PLACE	REGENT STREET	PACIFIC AVE
ELECTRIC COMPANY	ELECTRIC COMPANY	OXFORD STREET	NORTH CAROLINA AVE
WHITEHALL	STATES AVE	COMMUNITY CHEST	COMMUNITY CHEST
NORTHUMBERL'D AVE	VIRGINIA AVE	BOND STREET	PENNSYLVANIA AVE
MARYLEBONE STN	PENNSYLVANIA RLRD	LIVERPOOL ST STN	SHORT LINE
BOW STREET	ST JAMES PLACE	CHANCE	CHANCE
COMMUNITY CHEST	COMMUNITY CHEST	PARK LANE	PARK PLACE
MARLBOROUGH ST	TENNESSEE AVE	SUPER TAX	LUXURY TAX
VINE STREET	NEW YORK AVE	MAYFAIR	BOARDWALK

probabilities, and we are ignoring instructions from Chance and Community Chest, it is legitimate to break the game's moves into two steps.

Figure 35 shows a mathematician's view of the game board, interpreted in this manner. The lines join each square (represented by a blob) to each of the next six in the clockwise direction. I shall use this diagram to answer the simplest question about the ultimate fairness of

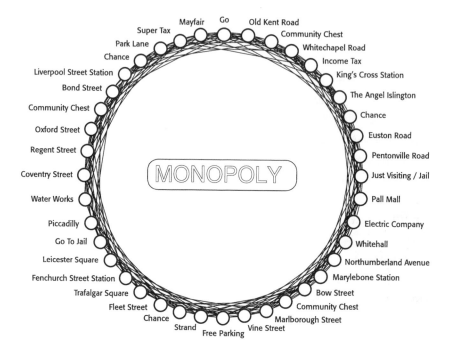

Figure 35 A mathematician's view of the Monopoly board (simplified version).

the game: after a large number of throws of the dice, do all squares become equally probable? We've just seen that this is *not* true after one player's turn (two throws, one ghost plus one real). And it's not true in the real game either, even for a large number of throws, as we'll see in the next chapter.

For convenience we number the squares from 0 to 39, with 0 = Go. Then square 40 'wraps round' to square 0, and we can think of the numbers as being counted modulo 40 – meaning that anything larger than 39 can be replaced by its remainder on division by 40. Now we imagine a single player making repeated throws of a single die, and moving accordingly, over and over again. The key question is: what is the probability of landing on a given square after a given number of throws? We would hope that when the number of throws becomes large, this probability becomes very close to 1/40, for *any* of the 40 squares. That is, they should all become (nearly!) equally likely.

The way to calculate these probabilities is to see how the distribution of probabilities 'flows' as more and more dice are tossed. Each distribution can be represented by a sequence of 40 numbers, the probabilities of landing on squares 0, 1, 2, ... , 39, respectively. At the start of the game, the player is on square 0 (Go) with probability 1 (that is, always). So the distribution of probabilities looks like

$$1, 0, 0, 0, ... , 0$$

with 1 on square 0 and 39 0s on the others.

After a single (ghost) toss the distribution becomes

$$0, 1/6, 1/6, 1/6, 1/6, 1/6, 1/6, 0, ... , 0.$$

That is, the probability of landing on each of squares 1, 2, 3, 4, 5, 6 is 1/6, and you can't get to any of the others.

Notice that the total probability of 1, originally concentrated on square 0, has been split into six equal parts, and these have been moved to the squares 1, 2, 3, 4, 5, and 6 units further along. This is a completely general procedure. At each toss of the die, the probability on a given square is divided by 6, and these six equal parts flow on to each of the next six squares in the clockwise direction. So on the next throw the 1/6 on square 1 is redistributed like this:

$$0, 0, 1/36, 1/36, 1/36, 1/36, 1/36, 1/36, 0, ... , 0.$$

The 1/6s on squares 2–6 are similarly redistributed, but shifted along one step each time:

$$0, 0, 0, 1/36, 1/36, 1/36, 1/36, 1/36, 1/36, 0, ... , 0.$$
$$0, 0, 0, 0, 1/36, 1/36, 1/36, 1/36, 1/36, 1/36, 0, ... , 0.$$
$$0, 0, 0, 0, 0, 1/36, 1/36, 1/36, 1/36, 1/36, 1/36, 0, ... , 0.$$
$$0, 0, 0, 0, 0, 0, 1/36, 1/36, 1/36, 1/36, 1/36, 1/36, 0, ... , 0.$$
$$0, 0, 0, 0, 0, 0, 0, 1/36, 1/36, 1/36, 1/36, 1/36, 1/36, 0, ... , 0.$$

Finally we add up the probabilities that have landed on each particular square. For example, square 6 (the seventh term in each sequence) acquires 1/36 from each of the first five sequences, but 0 from the last one, so the total is 5/36. The final result is

0, 0, 1/36, 2/36, 3/36, 4/36, 5/36, 6/36, 5/36, 4/36, 3/36, 2/36, 1/36, 0, ... , 0.

This tallies with our expectations for tossing two dice.

But now we can continue. On the third (ghost) throw, we multiply every term in the sequence we've just calculated by 1/6, and shift it up 1, 2, 3, 4, 5, and 6 terms. Then we add the numbers on each square together to get the probability distribution after three throws. And so on. When doing this it is crucial to remember that anything that flows off the end of square 39 gets wrapped round to the start again.

It's easy to write a computer program to calculate these probability distributions one by one. The results are represented graphically in Figure 36, starting with the 'triangular' distribution obtained on the second throw. On each subsequent throw the probability graph moves one step forward in the figure. You can see that the probability peak moves several squares to the right at each step (in fact on average it

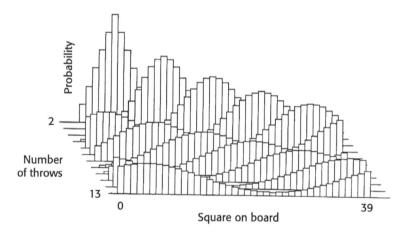

Figure 36 The distribution of probabilities over the 40 squares, and how it changes at each successive throw of the die. The height of each bar shows the probability of landing on the corresponding square; the graphs for throws 2–13 are stacked in three dimensions from back to front.

moves 3.5 squares, the mean value of the numbers 1, 2, 3, 4, 5, 6). The initial sharp triangular shape becomes more rounded and broadens out. Even after the first 13 throws the distribution is by no means uniform, but if the trend shown in the figure continues then eventually the peak will flatten out completely and all of the values will be pretty much the same. If you continue the computer simulation, you will find that this is indeed so. But it would be nice to gain some insight into *why* the simulations do this.

One approach is to appeal to physical intuition. Go back to Figure 35 and imagine that the probabilities are represented by an electric charge. At the start of the game, a total charge of 1 is concentrated on square 0. At each subsequent throw of the die, the charge on each square splits into six equal parts, and these parts flow along the lines of the figure to each of the next six squares. We would expect that eventually such a flow of charge would tend towards a unique steady state – and we can work out what that steady state must be by appealing to the symmetry of the picture. We have a polygon with 40 vertices, and the flow lines look exactly the same no matter which vertex you focus on. That is, the whole picture is symmetric under rotations of the polygon. Therefore the unique steady state must also be symmetric under all rotations. But this means that the steady state charge on each vertex is the same, and since the total charge is conserved, we must end up with charge 1/40 at each vertex. Reinterpreting charge as probability, we find that the distribution settles down towards a steady state in which each square carries probability 1/40.

But this argument, plausible though it may be, is not a proof. For that, we need Markov's theory, which provides a systematic method to track the probability flow. It begins by writing down the 'transition matrix' for the network shown in Figure 35, which is a 40×40 square table with rows and columns numbered 0–39. The entry in row r and column c of this table is the probability of moving, in one step, from square r to square c. This is 1/6 if $c = r + 1, r + 2, \dots , r + 6$ (modulo 40), and 0 otherwise. Call this transition matrix M.

There is then a technical calculation, carried out using M, and summarized in the box, whose result is that in the long run the probability distribution does indeed get as close as we wish to

$1/40, 1/40, /401, 1/40, \dots, 1/40.$

That is, the probability of landing on any given square is the same. So, with a little help from Markov, we can *prove* that a game as complicated as Monopoly is fair, in the sense that – in the long run – no particular square is more or less likely to be landed on. Of course, the first player still has a small advantage, but this is mitigated by the finiteness of their bank balance.

Like I said, this equality of probabilities holds in the idealized version of the game. As we'll see in the next chapter, it most assuredly does not in the real game. In particular, in the real game the 'Go to Jail' square produces a skewed probability distribution. In fact, the Jail square is the one most often visited – an interesting but probably unintentional comment on the business world – with a probability of 5.89% compared to the 'equidistributed' value of 2.5% (or 2.44% if 'Just Visiting' and 'Jail' are distinguished, which seems sensible, both within the business world and without it). The next most likely square is Trafalgar Square, probability 3.18%. The square visited least often is the third Chance square round from 'Go', probability 0.871, apart from 'Go to Jail' which is not actually *visited* at all because you're off to the pokey.

You might like to try the same kind of analysis for Snakes and Ladders, assuming that as soon as you land on a snake you slide down to its tail, as soon as you land on a ladder you clamber up to the top, and when you get to the end you stay there. Now the long-term distribution does *not* have equal probabilities. What do you think it is?

And here's a trickier problem, which I suggest you attack by computer simulation unless you're a real whiz with matrix algebra. What is the long-term probability distribution on the squares of a Snakes and Ladders board if you wrap the END square round to the start and keep going?

MARKOV'S MATRIX MAGIC

Let M be the transition matrix. The first step is to calculate a set of 40 numbers called the 'eigenvalues' of M. (A number m is an eigenvalue of M if you can write 40 numbers on the 40 vertices of the network so that when you split each into six and let it flow along the six clockwise lines emanating from that vertex, the resulting numbers are exactly m times the size of the numbers you started with. Phew. It's simpler expressed in symbols, really: $Mv = mv$ for some v.) But there's a twist: those numbers need no longer be probabilities – real numbers between 0 and 1 – but complex numbers, expressible using the number $i = \sqrt{-1}$.

The sequence formed by these 40 numbers, by the way, is called an 'eigenvector'.

Now, says Andrey Andreyevich, all you have to do is find the *biggest* eigenvalue among the 40 you've calculated. Then the probability distribution, in the long run, will be approximated as closely as you wish by the corresponding eigenvector – 'normalized' so that its entries add up to 1, like genuine probabilities should. (This just means that you divide every entry by the total.)

Because of the rotational symmetry of Figure 35, it is actually not hard to find the eigenvalues and eigenvectors. In particular, one eigenvector looks like

1/40, 1/40, /401, 1/40, ... , 1/40

with all 40 entries equal to 1/40. What is its eigenvalue? Well, suppose you start from this distribution, split each 1/40 into six equal pieces of size 1/240, and shove them along their six clockwise lines. Each vertex receives exactly six contributions: one from each of the six preceding vertices. So it ends up with $6 \times 1/240 = 1/40$ again. This is what an eigenvector *should* do, and in this case the eigenvalue is 1.

I won't tell you the other 39 eigenvalues, whose expressions are beautiful to mathematicians (only). But a calculation shows that they are all smaller (in absolute value, for purists) than 1. In fact the next largest has absolute value 0.964. So 1 is the largest eigenvalue, and its eigenvector

1/40, 1/40, /401, 1/40, ... , 1/40

does indeed represent the long-term state of the probability distribution.

10

Monopoly Revisited

Now for the real game of Monopoly®, which is far more complicated – so much so that even the Markov chain method cannot capture every single nuance. But it can show that the most frequently visited square is 'Jail', with a probability roughly double that of any other square. And the least likely destination is the third Chance square. The real fun starts when you try to model property-dealing mathematically. When, and where, should you build a house or a hotel?

In the previous chapter we looked at a simplified mathematical model of the game of Monopoly, asking whether the bunching of probabilities that is produced by everyone starting from the 'Go' square eventually evens out. The analysis depended on several simplifying assumptions, ignoring the result of throwing a double, the effect of the 'Go to Jail' square, Chance cards, and so on.

Of course, this is not an analysis of real Monopoly, and was never intended to be when I wrote the column concerned in April 1996. That didn't stop irate Monopoly fans writing in to express their outrage, with varying degrees of (im)politeness, but the advantage was that the more perspicacious ones sent me their own analyses of the real game. Which I now report for everyone's edification, as I did in October of the same year.

You will see shortly that the methods I applied to my simplified version can easily be modified to encompass the full set of rules. So why didn't I just do that? The real game is a bit too messy to give a good feel for the mathematics that I'm using to analyse it, so I thought it would be better to warm up on something simpler, but related. Recall the basic idea: to represent the game as a Markov chain, a matrix of transition probabilities, and then to compute the 'eigenvalues' and 'eigenvectors' of that matrix. Markov's theory then implies that the long-term probabilities of landing on any given square are given by the eigenvector with the largest eigenvalue.

The simplified model has a beautiful mathematical symmetry, which makes it possible to calculate the eigenvalues and eigenvectors *exactly*.

I didn't show you the formulas, but the symmetry did let me illustrate one aspect of the theory, by showing that the vector whose entries are all 1/40 really is an eigenvector, with eigenvalue one. I then asserted that all the remaining eigenvalues are less than one (which follows from the exact formula), thereby giving you the bones of a proof that in the simplified model the long-term probabilities are the same for all squares.

Is it legitimate to make that kind of sweeping simplification? In general, my main aim is to show you that mathematics is *interesting*. Utility is a secondary virtue, so many practical issues are often disregarded. For example Chapter 4 on Murphy's Law assumes that toast has zero thickness. Real toast has a definite thickness, but that doesn't imply that mathematicians don't know what toast looks like. It just means that on this occasion they chose to disregard that complicating factor. And so it is reasonable to disregard some of the rules of Monopoly, to keep the mathematical story simple and to illustrate the role of symmetry.

But it isn't reasonable to disregard the rules if you are trying to understand *real* Monopoly, or if you are trying to illustrate the utility of mathematics by applying it to that game. Which we shall now do, using the same principles but doing an awful lot more work. The simplified description set up all of those principles, in a case where they could be understood without too many technicalities.

Recall that the problem is to calculate the *probability* of landing on any particular square, after the game has been played for many rounds and the probabilities have settled down to a 'steady state'. Again we use the Markov chain method, but now the matrix of transition probabilities is not as pretty, and instead of working out an exact formula, we must resort to numerical approximations. Most correspondents sensibly solved the problem using computer algebra; some simulated the game for a huge number of moves and obtained empirical estimates.

The most extensive analyses came from William Butler of Portsmouth RI; Thomas H. Friddell, a Boeing engineer from Maple Valley

WA; and Stephen Abbot in the Mathematics Department at St. Olaf College, Northfield MN, who sent some joint work with his colleague Matt Richey. Butler wrote a Pascal program, Friddell used Mathcad, and Abbot used Maple. The discussion that follows is a synthesis of their results. All models of Monopoly make some kind of assumption about the degree of detail incorporated into the model, and there were insignificant differences in the assumptions made by various correspondents: I shall ignore these.

The first modification of my simplified model is to take full account of the rules for throwing the dice. A pair of dice is thrown, and if the result is a double, the player throws again; however three consecutive doubles lands them in jail. The throw of the dice is a tiny Markov chain in its own right, and can be solved by the usual method. The result is Figure 37, a graph of the probability of moving any given distance from the current position. Notice that the most likely distance is 7, but that it is possible to move up to 35 squares (by throwing 6:6, 6:6, 6:5). However, the probabilities of moving more than 29 squares are so small that they fail to show up on the graph.

Next the effect of the 'Go to Jail' square must be included. The jail rules, incidentally, pose a problem, because players can either elect to

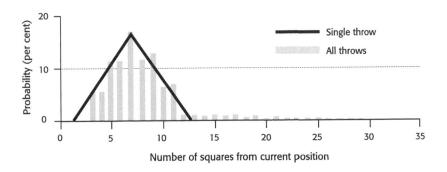

Figure 37 Probabilities of moving a given number of squares on one turn, taking rules concerning doubles into account.

buy their way out of jail or they can stay in and try to throw doubles. (Or at later stages when jail becomes a refuge from high rents they can stay in and hope *not* to throw doubles!) The probabilities associated with this choice depend upon the player's psychology and bankroll, so the process is non-Markovian. Most correspondents got round this by assuming that the player did not buy their way out. Then 'Jail' becomes not so much a single square as a Markov subprocess – conceptually a series of three squares where players move from 'Just in Jail' to 'In Jail One Turn Already' to 'Must Come Out Next Turn'. Of course the 'Go to Jail' square itself has probability zero because nobody actually *occupies* that square.

The next step is to modify the matrix of transition probabilities to take account of the effect of 'Chance' and 'Community Chest' cards, which may send a player to Jail or to some other position on the board. This can be done quite simply, by counting what proportion of these cards send the player to any given square.

Having set up an accurate transition matrix, the steady state probabilities can be worked out either by a numerical computation of eigenvalues and eigenvectors, or by calculating the effect of making a large number of moves in succession by forming powers M^2, M^3, ... of the transition matrix M. Mathematically these two methods are equivalent, thanks to Markov's general theorem which relates the powers of M to the eigenvector with largest eigenvalue.

The probabilities of occupying different squares, expressed as a percentage and accurate to nine decimal places, are shown in Table 3 and graphed in Figure 38. The most dramatic feature is that players are almost twice as likely to occupy the 'Jail' square (5.89%) as any other, the next most likely being Trafalgar Square (3.18%). Of the railroads, Fenchurch Street is occupied most often (3.06%) with King's Cross (2.99%) and Marylebone (2.91%) just behind; however the probability of occupying Liverpool Street is much less (2.44%). The reason is that unlike the others it does not feature on a Chance card. Among the utilities, Water Works (2.81%) wins out, with Electric Company (2.62%)

Table 3 Steady state probability of occupying any given square

Square	Probability (%)	Square	Probability (%)
GO	3.113802817	FREE PARKING	2.874825933
OLD KENT ROAD	2.152421585	STRAND	2.830354362
COMMUNITY CHEST	1.889769064	CHANCE	1.047696537
WHITECHAPEL ROAD	2.185791454	FLEET STREET	2.738576558
INCOME TAX	2.350777226	TRAFALGAR SQUARE	3.187794862
KINGS CROSS STN	2.993126856	FENCHURCH ST STN	3.063696501
THE ANGEL ISLINGTON	2.285359460	LEICESTER SQUARE	2.706510944
CHANCE	0.8760265658	COVENTRY STREET	2.679312587
EUSTON ROAD	2.347010651	WATER WORKS	2.810736815
PENTONVILLE ROAD	2.330647102	PICCADILLY	2.591184852
JUST VISITING/JAIL	5.896419869	GO TO JAIL	0
PALL MALL	2.735990819	REGENT STREET	2.686591663
ELECTRIC COMPANY	2.627460909	OXFORD STREET	2.633846362
WHITEHALL	2.385532814	COMMUNITY CHEST	2.376569966
NORTHUMBERL'D AVE	2.467374766	BOND STREET	2.510469546
MARYLEBONE STN	2.918611720	LIVERPOOL ST STN	2.445895703
BOW STREET	2.776751033	CHANCE	0.8715029109
COMMUNITY CHEST	2.571806811	PARK LANE	2.202178226
MARLBOROUGH ST	2.916516994	SUPER TAX	2.193106960
VINE STREET	3.071024294	MAYFAIR	2.646925903

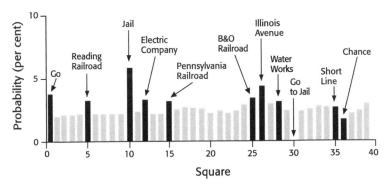

Figure 38 The steady state probability distribution: squares on the board are numbered 1–40 and the height of each bar is the long-term probability of occupying that square. See Table 2 (p. 99) for British names.

marginally less probable. Go (3.11%) is the third most likely square, and the third Chance square (0.87%) is the least likely – except for Go to Jail (0% occupation by logical necessity).

Friddell went further and analysed Monopoly's property market, which is what really makes the game interesting. His aim was to find the break-even point for buying houses – the stage at which income starts to exceed costs – and to find the best strategies for buying houses and hotels. The exigencies of the property market depend upon the number of players, and also which version of the rules is being played. Assuming that houses can be bought from the start – as is the case in the 'Short Game', a number of general principles emerge:

- Although it costs more to buy houses early, the break-even point will be reached more quickly if you do.

- With two houses or fewer it typically takes around 20 moves or more to break even. Three houses produces a definite improvement.

- Between Go and Fleet Street the property square that offered the quickest break-even point for three houses is Vine Street, which breaks even in about 10 turns.

Properties beyond Fleet Street were not evaluated: Friddell says he stopped there because he never expected to publish his results. Here's a sample of Friddell's property analysis techniques, applied to Vine Street. Figure 39 shows the probability of a player occupying that property on a given turn. Figure 40 shows how the overall value of the property (negative values are capital outlay) depends on the number of turns since it was acquired. There are six graphs, depending on how many houses/hotels are in place.

Many other readers contributed interesting observations and I can mention only a few. Simulations by Earl A. Paddon of Maryland Heights MO and calculations by David Weiblen of Freetown CT confirmed the pattern of probabilities. Weiblen points out that these

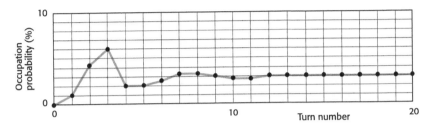

Figure 39 The probability of occupying Vine Street on a given turn.

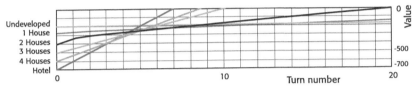

Figure 40 Average cash-flow for properties on Vine Street.

probabilities do not really affect how 'fair' the game is, because all players face the same situation. (I was thinking of 'fair' in the sense of 'fair coin' – that is, unbiased – but it's worth clarifying the meaning and I agree entirely.) Developing this point, he notes that 'If the rewards for landing on low-probability squares were out of proportion to that lowered probability, then there would be a problem. When out of sheer luck, a player in a game gets a big advantage, the game is unfair.' But he concludes that Monopoly is not unfair in that manner.

Bruce Moskowitz of East Setauket NY added yet another dimension to the analysis by remarking that 'In my youth I played Monopoly many times with my brothers and friends, and it was common knowledge that the tan-coloured properties, St. James Place, Tennessee Avenue and New York Avenue [Bow Street, Marlborough Street, and Vine Street], are especially valuable since there is a relatively high probability of landing on one of them when leaving Jail.' This analysis receives theoretical confirmation, for all three of these properties figure among the top 12 in the list of probabilities. Several other properties are also more likely to be landed on than normal, for other reasons.

Jonathan Simon of Cambridge MA chided me for suggesting that cheap properties were put near the start to help even out the game. 'Monopoly was ... created during the Great depression by a single designer, Charles Darrow, with lots of presumably unwelcome time on his hands. Under the trappings of wealth, the illustrated fat and rich men, it is (slyly) a poor man's game. In virtually all Monopoly contests ... the "cheap" properties turn out to be the most vital to Monopolize. Given the limited bankroll of each player, it is the New York, Virginia, Connecticut [Vine Street, Northumberland Avenue, Pentonville Road], and yes, even the Baltic [Whitechapel Road] groups that can feasibly be built up early enough in the game to tilt the balance. The 'lucrative' properties ... are expensive to own and prohibitively expensive to build without a source of income provided by ownership of a cheap group with houses.'

Point taken, though I would still argue that putting a lucrative property on the first half of the board would definitely be unfair, by Weiblen's criterion that no player should gain a big advantage purely by chance. And I'm not convinced that buying up lots of cheap properties and renting them out is a poor man's strategy!

11

A Guide to Computer Dating

If you think our usual calendar, with its funny rules for leap years, is complicated, what about the old Hindu calendar, based on a cycle exactly 1577,917,500 days long? Or the Chinese calendar, where a year can contain 12 or 13 months? Why are there so many calendars, and why is every one of them a compromise? Because a calendar that matches the true cycles of the heavens is a mathematical impossibility.

In 46 BC the Roman calendar was getting out of synch with the seasons. On the advice of the Greek astronomer Sosigenes, Julius Caesar introduced an extra day into every fourth 'leap' year to make the average length of the year 365¼ days. Misunderstanding the rule, his priests counted the fourth year of one cycle as the first in the next, so every *third* year became a leap year. The mistake wasn't fully sorted out for 50 years. For all our supposed sophistication, we have not learned from Caesar's priests, as is shown by the tale of the 'Millennium Bug', in which it turned out that the majority of the world's computers were incapable of coping with any date after 31 December 1999, interpreting 2000 as 1900. In fact, most of them – the ones using the most popular operating system – couldn't even cope successfully with the correct rule for leap years. In the event, planes did not fall out of the sky at one minute after midnight on 31 December 1999, as widely predicted. Oh, and the actual millennium started on 1 January 2001, not 2000, because there was never a year zero, but most people dislike being reminded of that.

We need not make the same mistake again. About ten years ago Nachum Dershowitz and Edward M. Reingold of the Department of Computer Science, University of Illinois at Urbana-Champaign, decided to develop calendar and diary features for the Unix-based editor GNU-Emacs. Out of this project grew a unique resource: computer code for converting dates from one calendric system to another. The 14 calendars they decided to cater for are the Gregorian, ISO, Julian, Coptic, Ethiopic, Islamic, Persian, Bahá'í, Hebrew, Mayan,

French Revolutionary, Chinese, old Hindu, and modern Hindu. Their book *Calendric Calculations* (Further Reading) is an absolute goldmine for chronologists.

Calendars vary from culture to culture because they are all attempts to perform the impossible: to rationalize the irrational. Our units for time are based on three distinct astronomical cycles: the day, month, and year. A normal 24-hour *mean solar day* is the period between successive occasions when the Sun is overhead. (One rotation on its axis, relative to the 'fixed stars', takes 23 hours 56 minutes and 4 seconds – but the Earth is also revolving around the Sun, and it takes a further four minutes for the extra rotation to compensate for the Sun's apparent slippage across the sky.) The period between successive new moons is the *mean synodic month*, which lasts 29.530588853 days. The period required for the Sun to return to the same position in its apparent path is the *mean tropical year* of 365.242189 days.

If the month were 29.5 days and the year 365.25, then the Moon would repeat its motion exactly every 59 days (2 × 29.5) and the Sun every 1461 days (4 × 365.25). So every 86,199 days (59 × 1461) the system of Earth, Moon, and Sun would return to precisely the same relative position. A calendar with an 86,199-day cycle would remain in step forever – ignoring slow changes to the lengths of the day, month, and year caused by forces such as tidal friction. Unfortunately for calendar designers, the ratios between days, months, and years behave like irrational numbers: they are not expressible as exact fractions. (At least, not using smallish integers: 29,530,588,853/1000,000,000 would lead to an impractically long cycle.) So in practice the lunar and solar cycles never return to *exactly* the same state at exactly the same time of day.

The day is central to timekeeping because of the day–night cycle, the lunar month is important for many cultures for religious reasons, and the year determines the cycle of the seasons, so a comprehensive calendar has to include all three. In practice most cultures either plump for a solar calendar, and fudge the months, or a lunar calendar, and ignore problems with the seasons. Whatever the choice, the calendar-designer must find practical ways to deal with small cumulative errors, hence the complicated paraphernalia of leap days, months of variable length ('Thirty days hath September ...'), and so on. To find out just how complex it can get you should either consult a copy of *Calendrical Calculations* or visit its home page:

http://emr.cs.iit.edu/~reingold/calendar-book/second-edition

Here I shall try to convey some of its unique and fascinating flavour, but skipping many fine points.

The simplest calendric system would ignore years and months and number consecutive days, choosing some convenient 'epoch' (start day). Astronomers use one such system, the Julian day, but Dershowitz and Reingold prefer an invention of their own: the 'fixed date' or 'rata die', abbreviated to RD. Day 1 of the RD system is January 1 in year 1 of the Gregorian calendar, the calendar we now use. There was no actual year 1 in the Gregorian calendar, since that was introduced in 1582 by Pope Gregory XIII, so we extrapolate backwards. That particular day was a Monday, which is convenient since we can take day 0 to be the previous Sunday and number the days of the week from 0–6 starting on Sunday. *Calendrical Calculations* uses the RD value as a common reference system: for example, to convert a date in the Hebrew calendar to one in the Chinese, you convert from Hebrew to RD and then from RD to Chinese. This way only 28 conversion functions (one each way for each of the 14 calendars) are needed.

Here are two simple warm-up problems which exemplify the type of mathematics required:

1. What day of the week will 1000,000 RD be?
2. How many complete mean tropical years will elapse between 0 and 1000,000 RD?

To answer question 1, observe that the days of the week form a repeating cycle of length seven, 'wrapping round' as shown in Figure 41. Therefore any RD that is a multiple of 7 must be a Sunday, any that leaves remainder 1 on division by 7 is a Monday, and so on. We say that

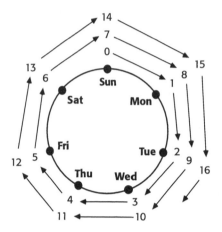

Figure 41

Finding the day of the week by counting modulo 7.

the day number is the RD number *modulo* 7. 'Modulo' is Latin for 'to the modulus': x mod 7 means 'find the remainder upon dividing x by 7'. Since $1{,}000{,}000 = 7 \times 142{,}857 + 1$, this remainder is 1 when $x = 1{,}000{,}000$, so 1,000,000 RD is a Monday.

To answer the second question, divide 1,000,000 by 365.242189 to get 2737.9094. This tells us that 1,000,000 RD occurs 2737 complete (mean tropical) years after 0 RD, a number that we find by omitting everything after the decimal point. Mathematically this is performed by the 'floor function' $\lfloor x \rfloor$, which is the greatest integer less than or equal to x.

Now consider converting a Gregorian date, such as December 25 1996, to its RD value. Recall Pope Gregory's leap year rule, which makes the average length of a year more accurate: multiples of 4 have an extra day on 29 February, unless they are multiples of 100, however multiples of 400 are also leap years. Dershowitz and Reingold show that this leads to the calculation rule in the box. For example, let $M = 12$, $D = 25$, $Y = 2100$. Then (a) = 766135, (b) = 524 − 20 + 5 = 509, (c) = 336, (d) = −1, and (e) = 25. So the RD value of December 25 1996 is $766{,}135 + 509 + 336 - 1 + 25 = 767004$. As a simple application, the day of the week is therefore 767,004 mod 7 = 0, so Christmas 1996 happens on a Sunday.

●●●

TO FIND THE RD VALUE OF MONTH *M*, DAY *D* AND YEAR *Y* GREGORIAN

Compute:

(a) $365(Y - 1)$

(b) $\lfloor (Y-1)/4 \rfloor - \lfloor (Y-1)/100 \rfloor + \lfloor (Y-1)/400 \rfloor$

(c) $\lfloor (367M - 362)/12 \rfloor$

(d) 0 if $M \leq 2$, −1 if $M > 2$ and Y is a leap year, and −2 otherwise

(e) D

and add them together.

The calculation has the following interpretation: (a) is the number of non-leap days in prior years. (b) is the number of leap days in prior years (one every fourth, except that every 100th is omitted, but you put back every 400th). (c) is a cunning formula for the number of days in prior months of year Y, based on the assumption that February has 30 days, which it doesn't – hence the correction term (d). In step (e) the number D is of course the number of days in the current month – the only days not yet counted.

●●

To see the complexity that the software in *Calendrical Calculations* handles with ease, consider the modern Persian calendar. It was adopted in 1925, but its epoch is March 19 622 AD – the vernal equinox prior to the epoch of the Islamic calendar. It is closely based on the more ancient Jalalai calendar invented by a committee of astronomers that included Omar Khayyam. There are 12 months: the first six (Fravardin, Ordibehest, Xordad, Tir, Mordad, Sahrivar) have 31 days, the next five (Mehr, Aban, Azar, Dey, Bahman) have 30, and the last (Esfand) has 29 in an ordinary year and 30 in a leap year. The leap year pattern, taken unchanged from the Jalali calendar, is highly intricate. It follows a cycle of 2820 years, containing 683 leap years. The 2820 years are divided into twenty-one 128-year subcycles, followed by one of 132. Each 128-year subcycle is divided into sub-subcycles of lengths 29 + 33 + 33 + 33; whereas the 132-year subcycle is divided as 29 + 33 + 33 + 37.

Finally, in each sub-subcycle years 5, 9, 13, and so on, going up in fours, are leap years. The Persian calendar is in error by 1.7 minutes at the end of one 2820-year cycle, so it would take 2.39 million years to slip a day relative to the true astronomical cycles!

The old Hindu lunisolar calendar follows a very different pattern. The months follow the Moon's phases closely, and an additional leap month is 'intercalated' to keep the months in step with the solar year. Unlike most such systems, however, the cycle of intercalation does not follow a fixed, simple pattern. The overall structure involves a cycle lasting 1577,917,500 days. The 'year' (strictly the *arya sidereal year*) is one 4320,000th of this, or 365.258 days. The *solar month* is one-twelfth of a year, and there are 12 named months. The lunar month is one 53,433,336th of the 1577,917,500-day cycle, equal to 29.531 days. The basic idea is to run both month-lengths simultaneously. Usually a lunar month overlaps a boundary between solar months, but every so often a lunar month is completely contained in a solar month. In that case, it is considered to be a lunar leap month, and its name is also given to the following lunar month (Figure 42).

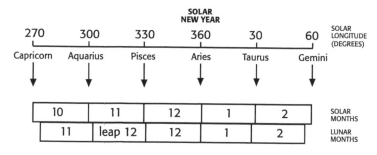

Figure 42 The Hindu lunisolar calendar. A leap month is added whenever a lunar month fits inside a solar one.

Finally, let's take a look at the Chinese calendar, which is based on astronomical events, not arithmetical rules. The Chinese calendar has been reformed at least 50 times: the version implemented in *Calendric Calculations* is the most recent, dating from 1645, the second year of the

Qing dynasty. Months are lunar, beginning at the day of the new moon, and years contain either 12 or 13 months. The arrangement of the months, however, depends upon the passage of the Sun through the signs of the zodiac. The solar year is divided into twelve major solar terms called *zhongqi* and twelve minor solar terms called *jieqi*. Each term corresponds to a 15° segment of solar longitude, the major ones starting at multiples of 30° and the minor ones in the gaps between those. These terms occupy roughly the same period in any year (Figure 43). The basic rule that determines the calendar is that the winter

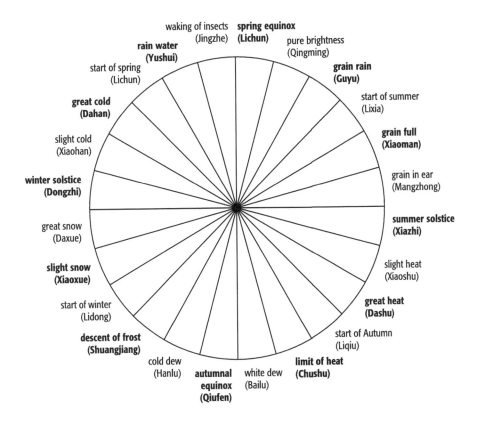

Figure 43 Solar terms of the Chinese year (major terms in boldface).

solstice *always* occurs during the eleventh month of the year. In a year that contains only 12 complete lunar months, therefore, the months are always numbered 12, 1, 2, 3, 4, 5, 6, 7, 8, 9, 10, 11. In a year that contains 13, however, one of the numbers is duplicated in a leap-month. Which? It is the first month that does not contain a major solar term. (Since there are 13 lunar months and only 12 major solar terms, at least one month must fail to contain a major solar term – an application of the so-called 'pigeonhole principle': if there are more holes than pigeons, then at least one hole must be pigeonless.)

Since present-day calendars are so complex, what of the future? Now the mathematics required is dynamics, plus the sciences of astronomy, physics, climatology ... All of the various astronomical cycles are slowly changing their lengths because of tidal gravitational forces. Moreover, there is the 'precession of the equinoxes', which is not steady, but has occasional glitches related to ice ages – so a future calendar must be linked to climate. In fact a future calendar must be interactive, adjusted according to what actually happens, not just based on preset rules, because astronomers Jack Wisdom (MIT) and Jacques Laskar (Bureau des Longitudes, Paris) have discovered that the motion of the solar system is chaotic, so if you set up a fixed calendar to keep pace with the seasons, the infamous 'butterfly effect' will cause it to drift away from reality. Independence Day in 10,000,000 AD may still be the fourth of July, but nobody can predict how many days from now that will be.

12

Dividing the Spoils

Dividing a cake fairly between two people is easy: 'I cut, you choose'. With more people, fair division is a more complicated task. And what we mean by 'fair' is also more complicated. You may feel that you have got your fair share – but you may also feel that someone else got more than their fair share. A way of dividing the cake that avoids this problem is 'envy-free'. Until recently, an envy-free method of division was known only for two or three people. But now ...

Artful Arthur tipped the sack out on to the kitchen table. Bent Bertha, Crooked Clare, and Dennis the Dodger stared, eyes popping, as wads of banknotes and heaps of jewellery tumbled out.

Now came the difficult bit: dividing the spoils. None of them trusted the others, and they were all determined that nobody else would get a bigger share than them. Arthur rapidly shared out the money, which was easy as long as everybody watched very carefully. Dividing the jewellery was going to be more difficult, because each of them would have their own views on what it was worth. Fortunately a lot of it was things like gold chains, which could be cut into pieces if necessary.

'We'd better get a move on,' said Clare nervously. 'The police won't be far behind us and we need to get the stuff hidden away.'

'I'll have that diamond tiara, then,' said Bertha, jamming it on her head. 'Then Clare can have the necklaces, and – '

'Those necklaces are rubbish!' yelled Clare. 'I want the emerald brooch and the – '

'Quiet!' shouted Dennis. 'You bunch of crooks will never agree if all you do is argue. What we need is a method, guaranteed to satisfy each of us that they've got their fair share.'

'Yeah.'

They stared at the heap of assorted jewellery.

Three hours later, they were still staring at it. 'What we really need,' said Dennis, is what mathematicians call a 'proportional envy-free allocation protocol'.'

'Yeah,' said Arthur. 'Sure. Uh, Dennis – what does that gibberish mean?'

'Envy,' Clare interrupted, 'is when one person resents the fact that somebody else has got something that they – '

'Envy I understand,' said Arthur. 'I mean the long words.'

'Oh. Really?'

'An allocation protocol,' said Dennis, before there was a fight, 'is a systematic method for dividing things between several people. It is proportional if at the end each person is satisfied that they have got at least their fair share; and it is envy-free if nobody thinks anybody else has got *more* than their fair share.'

'Isn't that the same thing?' asked Bertha.

'Not at all,' said Clare. 'Aren't I right, Dennis?'

He nodded. 'Envy-free protocols are always proportional, but proportional ones need not be envy-free.'

'Why not?'

'For instance,' said Dennis, 'suppose the three of you are sharing three items – a bracelet, a necklace, and some earrings, say. Then you might subjectively assess the proportions like this:

	Bracelet	Necklace	Earrings
ARTHUR	**40%**	50%	10%
BERTHA	30%	**50%**	20%
CLARE	30%	20%	**50%**

You get a proportional allocation if (see the boldface numbers) Arthur gets the bracelet, Bertha gets the necklace, and Clare gets the earrings. Each of them thinks their share is worth more than 33%. But Arthur still envies Bertha, because she's got the necklace, which he thinks is more valuable than the bracelet.'

As we shall see, envy-free protocols are much harder to obtain than proportional ones. You should also bear in mind that along with the protocol there is a set of strategies, one for each person, which guarantees that they will think they have achieved their goal. These strategies are not part of the protocol, and are often only tacit, but they are what guarantees it to be fair. We will often call the people 'players', because it helps to think of the whole process as a series of moves in a game, and it is customary to represent the valuables that are to be divided as a cake.

For two players, say Arthur and Bertha, there is a simple envy-free allocation protocol: 'I cut, you choose'. Arthur divides the spoils into two parts; Bertha chooses which of the two parts she wants (Figure 44).

Figure 44
The traditional method of dividing a cake between two people.

I CUT...   ...YOU CHOOSE

The strategies that go along with this protocol are:

Arthur's strategy: Divide the spoils 50/50 so that he cannot lose out no matter which part Bertha picks.

Bertha's strategy: Pick whichever part seems to her to be the bigger.

Arthur's strategy ensures that he gets a share that is at least 50% in his estimation – and so does Bertha's strategy. Therefore the allocation protocol is envy-free. And therefore proportional, of course.

Since it was Dennis's idea to find a protocol for sharing out the plunder, the other three crooks sent him off to the local university library to see what he could find. He soon came back with several books, which, it turned out, he'd stolen.

'Why didn't you just *borrow* them, Dennis?'

'Sorry, Bertha. Force of habit.'

Dennis had discovered that this area of mathematics came into being in wartime Poland, in the city of Lvov. In 1944, as the Russian army fought to reclaim Poland from the Germans, the mathematician Hugo Steinhaus sought distraction in a puzzle. He knew about the 'I cut, you choose' protocol for sharing a cake between two people, and he knew why the strategies associated with that protocol lead each player to believe that their share is at least half. The first player divides the cake into pieces which – in his view – are exactly equal. If the second player disagrees, she picks the piece she thinks is the bigger. Neither has any cause for complaint. If the first player is dissatisfied with the final outcome, he should have been more careful when making the first cut; if the second player is dissatisfied, she chose the wrong piece. Neither was forced at any stage to make a choice that they could deem unfair.

'Steinhaus,' said Dennis to Arthur, Bertha, and Clare, 'wondered whether you could divide a cake between three people – you three, say.'

'OK, said Arthur. Maybe I should cut the cake into three pieces – which I think are equal – and then Bertha gets to choose one, followed by Clare?'

'No,' protested Bertha, 'that doesn't work. Clare and I may both think that the same one of Arthur's pieces is larger than a third of the cake, while the other two are smaller. If so, I get that piece and Clare is dissatisfied.'

Dennis told them that Steinhaus had come up with a complicated series of nine steps whose outcome was that each player would be satisfied that they had received at least one third of the cake. That is, he found a proportional protocol, to be explained in a moment. Some terminology will be useful. Say that a player thinks a piece is 'fair' if its size is one third (or more), 'unfair' if they think its size is *less* than one third. To 'pass' at some step is to do nothing. For convenience we talk about the *size* of pieces, but what we really mean is their relative value as judged by the player concerned. All judgements are subjective, made by the person performing the action involved. Statements in brackets are not part of the protocol, but they explain why it works: they are comments on the strategies available to the players to ensure they get their fair share.

1. Arthur cuts the cake into three pieces (which he thinks are all fair, hence subjectively equal).

2. Bertha can either
   • pass (if she thinks at least two pieces are fair) or
   • label two pieces (which she thinks are unfair) as being 'bad'.

3. If Bertha passed, then Clare chooses a piece (which she thinks is fair). Then Bertha chooses a piece (which she thinks is fair). Finally Arthur takes the last piece.

4. If Bertha labelled two pieces as 'bad', then Clare is offered the same options as Bertha – pass, or label two pieces 'bad'. She takes no notice of Bertha's labels when choosing her own.

5. If Clare did nothing, then the players choose pieces in the order Bertha, Clare, Arthur (using the same strategy as in step 3.)

6. Otherwise both Bertha and Clare labelled two pieces as 'bad'. Therefore there must be at least one piece that they *both* consider 'bad'. Arthur takes that one. (He thinks *all* pieces are fair, so he can't complain.)

7. The other two pieces are reassembled. (Clare and Bertha both think the result is at least 2/3 of the cake.) Now Clare and Bertha play cut-and-choose to share what's left between themselves (thereby getting what they each judge to be a fair share).

'Wow,' said Arthur. 'That's heavy.'

'You wait,' said Dennis. 'You ain't seen nuthin' yet. The trouble is that Steinhaus's method, although proportional, is *not* envy-free. It's possible to find cases where each player thinks they have a fair share, but (say) Clare thinks that Bertha has a bigger share than Clare does.'

For example, suppose Bertha thinks Arthur's division is fair. Then the protocol stops after step 3, and both Arthur and Bertha consider all three pieces to be of size 1/3. Clare must think her own piece is size at least 1/3, so the allocation is proportional. But if Clare sees Arthur's piece as 1/6 and Bertha's as 1/2, then she will envy Bertha, because Bertha got first crack at a piece that Clare *thinks* is bigger than hers.

'Did Steinhaus find an envy-free protocol?' asked Clare.

'Nope,' said Dennis. 'He didn't worry about the envy-free question at all: that came rather later.' Arthur gritted his teeth and peered nervously out of the window. No sign of the police yet. 'What did concern him,' Dennis continued, 'was finding a protocol for proportional allocation among four or more people. And his friends Stefan Banach and B. Knaster soon came up with one.'

Suppose there are n players, and call them $P_1, P_2, \dots, P_n$. This time say that a player thinks a piece is fair if it has size $1/n$, and unfair if it is smaller. Then the Banach–Knaster protocol is:

1. P_1 cuts a (fair) piece C.

2. P_2 is offered a choice:
   - pass (if he thinks C is unfair), or
   - trim C (to create a fair piece which we continue to call C).
   Set the trimmings aside temporarily.

3. P_3 is offered the same choice with the new C; then P_4 is offered the same choice; and so on until every player except P_1 has had the opportunity to trim C if they wish.

4. If nobody trimmed C, it goes to P_1. If it was trimmed, the last player to trim it gets C. (And considers it fair.)

5. The rest of the cake, plus trimmings, is reassembled. The remaining $n-1$ players (who all consider that at least $(n-1)/n$ of the original cake now remains) repeat the same procedure.

6. This continues until only two players are left. Then they play cut-and-choose.

'Again, the Banach–Knaster protocol is proportional but not envy-free,' Dennis pointed out.

'Yeah,' said Bertha. 'It also *differs* from the Steinhaus protocol when $n = 3$. In fact, it's simpler.'

'Quite right,' said Dennis. 'That's because it involves a new idea, trimming. And Steinhaus's contribution had already introduced another important idea: divide up part of the cake, and then focus on what's left.'

Bertha leaned back in her chair and gazed at the ceiling. 'Of course, in our problem we don't necessarily agree with each other about what each item is worth.'

'Ah,' said Dennis. 'That was true for Steinhaus too. A key feature of this class of problems is that it is not necessary for the players to agree in their evaluations of the various bits and pieces of the cake. The possibility of subjective differences is a crucial feature. In fact, as Steinhaus noticed, the problem is generally *easier* if people disagree.'

'Why is that? It sounds most unlikely.'

'If I like the icing on the cake and you like the marzipan, Bertha, then we can both be satisfied rather easily.'

Arthur grunted to show he'd seen the point. 'Dennis, these protocol things are a bit finicky. The police might show up at any minute. Are there any other methods we could use?'

'Sure. There are abstract mathematical existence proofs, based upon the so-called Liapunov Convexity Theorem. These tell us that equitable divisions of the cake always exist, but they don't tell us how to find such a division. Then there are "moving knife" algorithms.'

The following, due to L. E. Dubins and Edwin Spanier in 1961, is a sample. A large knife is slowly moved across the cake, parallel to itself. Players shout 'cut!' as soon as they are willing to accept the resulting slice. However, the moving-knife method involves a potentially infinite number of decisions, because at each instant of time each player must decide yes/no. So it isn't a true algorithm. All the protocols discussed below involve a discrete sequence of decisions.

'So we're stuck with protocols,' said Arthur. 'If we wait infinitely long, the police will *definitely* get here.'

'I wouldn't bet on it,' muttered Clare under her breath.

Dennis kicked her under the table and carried on with his story. 'Yup. In the early 1960s an envy-free protocol for three players was found independently by John Selfridge and John Horton Conway.'

Their method circulated informally among aficionados of recreational mathematics, and eventually found its way into print in Martin Gardner's 'mathematical games' column in *Scientific American*. It goes like this:

1. Arthur cuts the cake into three (fair) pieces.

2. Bertha may either:
   - pass (if she thinks two or more pieces are tied for largest) or
   - trim (the largest) piece (to create such a tie). Any trimmings are called 'leftovers' and set aside.

3. Clare, Bertha, and Arthur, in that order, choose a piece (that they think is largest or tied largest). If Bertha did not pass in step 2 then she must choose the trimmed piece unless Clare chose it first.

   [At this stage, the part of the cake other than the leftovers has been divided into three pieces in an envy-free manner – a 'partial envy-free allocation'. This takes a little checking, but it's true!]

4. If Bertha passed at step 2 there are no leftovers and we are done. If not, either Bertha or Clare took the trimmed piece. Call this person the 'non-cutter', and the other one of the two the 'cutter'. The cutter divides the leftovers into three pieces (that she considers equal).

   [Arthur has an 'irrevocable advantage' over the non-cutter, in the following sense. The non-cutter received the trimmed piece, and even if she gets all the leftovers, Arthur still thinks she has no more than a fair share, because he thought the original pieces were all fair. So *however* the leftovers are now divided, Arthur will not envy the non-cutter.]

5. The three pieces of leftovers are chosen by the players in the order non-cutter, Arthur, cutter. (Each chooses the largest piece, or one tied for largest, among those available.)

[The non-cutter chooses from the leftovers first, so has no reason to be envious. Arthur does not envy the non-cutter because of his irrevocable advantage; he does not envy the cutter because he chooses before she does. The cutter can't envy anybody since she was the one who divided the leftovers.]

'That's fine,' said Arthur, whose patience was running rather thin by now, 'but there are four of us, not three.' He narrowed his eyes, stared at Dennis and pulled a large knife from his belt. 'Of course, we could fix that if we have to.'

'Won't be necessary, I assure you,' said Dennis hurriedly. 'It might have been a few years ago, because at this point the subject rather stagnated. It was known that for four or more players an envy-free allocation always *exists*, but nobody could come up with a protocol that would produce such an allocation in a finite number of steps.'

Clare leaned forward. 'So what happened?'

'The science writer Dominic Olivastro,' said Dennis. 'He wrote a survey of the subject for the magazine *The Sciences*. Steven Brams, a political scientist at New York University who has written books on Game Theory, read the article and was hooked. Brams had long been fascinated by political and economic problems of fair division – such as the partitioning of Germany between the Allies at the end of World War II. Here was the same question posed in a purely mathematical form.

'Brams started by looking for an envy-free protocol for three players, not realizing that one had already been found by Selfridge and Conway. His method amounted to the first three steps in theirs, a *partial* envy-free allocation. But instead of their intricate method for dividing the leftovers, Brams just used the same method all over again.'

'But that just creates more leftovers,' protested Bertha.

'Sure, but they're second-order leftovers, a lot smaller than the first ones. A third application of the method deals with them, and so on.'

'Is that really a protocol? It doesn't necessarily stop after finitely many steps.'

'Right. But it was simple and it worked.'

'Fair enough, I guess ...'

'You'll be satisfied by the eventual method, which *does* stop. Anyway, Brams, encouraged by his early success, moved on to the case of four people – and got stuck.'

Arthur fingered his knife meaningfully, and snarled. Dennis gulped and continued hastily. 'At that point he got in touch with Alan Taylor, a mathematician friend at Union College, Schenectady. Taylor thought about the problem while giving a final examination to his students, taking the same cavalier attitude that leftovers don't really matter much; and he solved it. His solution was weird, because the first step was to divide the cake into five pieces – even though there were only four players.'

'That's a curious piece of lateral thinking.'

'Right. Taylor admits that he has no idea where it came from.'

Here's Taylor's partial envy-free protocol for four players:

1. Arthur cuts the cake into five pieces (which he thinks are equal).

2. Bertha trims up to two pieces, if necessary, to create a three-way tie for largest (in her opinion) and sets the trimmings aside.

3. Clare then trims one piece, if necessary, to create a two-way tie for largest (in her opinion).

4. Denis chooses first, then Clare, then Bertha, and Arthur takes the last piece left. If either Clare or Bertha trimmed a piece, they must choose such a piece if it is available at their turn.

It is not hard to see that each player thinks their piece is at least tied for largest, so the allocation is envy-free.

'That's is all very well.' said Arthur. 'But it doesn't fully solve the problem, does it, Dennis?'.

'No. Selfridge and Conway's protocol is finite: it doesn't go on forever. But Brams's cavalier approach to leftovers – repeat indefinitely – does go on forever, so, strictly speaking, they had not found a genuine protocol. This didn't worry Brams much – in political science there are always loose ends, and a few tiny bits of cake missing hardly mattered – but it worried Taylor, the mathematician, a lot. So he beavered away for several months, assisted by two colleagues, William Zwicker and Fred Galvin, until he found a way to rearrange the sequence of choices so that the method always stopped, with not even a tiny crumb left over.'

Even for four players, the resulting protocol is extremely complex: it goes through 20 steps, one of which is a lengthy sequence of trim-and-choose decisions by various players in turn. Because of its complexity, the method is described in a separate box. At two stages the protocol involves picking a number that is related to various players' subjective estimates of the relative sizes of pieces, so the number of steps depends upon the precise preferences. Whatever the preferences are, the number of steps is finite, but it can be made as large as we please by setting initial preferences appropriately. Notice that unlike the partial allocation originally found by Taylor, it starts more conventionally with a division into four pieces. However, his new idea of using many more pieces than players turns up several times, and his original five-piece protocol shows up in its entirety in a lengthy sequence in which leftovers are repeatedly subdivided.

Arthur, Bertha, Clare, and Dennis began working their way through the Brams–Taylor protocol. After about two hours, the table was covered with sheets of paper and lengthy scribbled calculations. Arthur was staring at a single large emerald.

'I think we have to cut this into twelve pieces,' he said, glaring at Dennis the Dodger.

'Yeah, well – the protocol's fine for cakes, of course, and anything else that can easily be subdivided as far as you want.' He saw the look on Arthur's face. 'Um, the problem with bits that can't be subdivided is much harder, and it may not have any solutions at – ' He squeaked in fear. 'Um, Arthur, Freddy the Fence could chop up that emerald faster than you could say boo to a goo – '

'I don't want Freddy the Fence involved in this,' said Arthur. 'You've wasted a great deal of precious time, Dennis, and I don't like that. The police may be on our trail by now, and if they are, it's all your – '

He stopped. In the distance, they heard the wail of a siren.

It grew louder.

Arthur turned to Bertha and Clare. 'Time for a disproportionate mercy-free allocation, I think.' He pointed his knife at Dennis. 'I cut?'

'Yeah. And we'll choose,' said Bertha.

THE BRAMS–TAYLOR PROTOCOL FOR FOUR PLAYERS

1. Bertha cuts the cake into four (fair) pieces and hands one to each player.

2. Arthur, Clare, and Denis are asked in turn whether they object to this allocation (which they do if they envy another player).

3. If nobody objects, stop.

4. Otherwise, work with the first player to object. Suppose (by choice of names) that it is Arthur. Arthur chooses a piece that he envies and calls it A; his original piece is called B. Having chosen A and B, the rest of the cake is reassembled for later consideration.

5. Arthur names a whole number $p \geq 10$ (This p is chosen to have the following curious property. Suppose that A is subdivided, in any manner, into p pieces. Then Arthur prefers A to B, even if the 7 smallest pieces are removed. He can achieve this by taking $p > 7a(a-b)$ where a is his evaluation of A, and b his evaluation of B.)

6. Bertha divides each of A and B into p pieces (which she thinks are equal).

7   Arthur chooses (the smallest) three pieces of B and names them S_1, S_2, S_3.
    He also either:
    - Chooses (the largest) three pieces from A (if he thinks these are all strictly bigger than all the Ss) and trims at most two of them (to the size of the smallest among those three), or
    - Subdivides (the largest) one of the parts of A into three (equal) pieces.

    Whichever he does, he names these pieces T_1, T_2, T_3.

8. Clare takes the six Ss and Ts, and either
    - passes (if she thinks there is already a two-way tie for largest)
    - trims (the largest) one (to create such a tie).

9. Denis, Clare, Bertha, and Arthur, in that order, choose a piece from the six Ss and Ts, modified as in step 8 (that they think are largest or tied largest). Clare must take the piece she trimmed if it is available. Bertha must choose an S; Arthur must choose a T.

    By this stage we have achieved an envy-free partial allocation, with a lot of leftovers, in which Arthur thinks his piece is strictly larger than Bertha's – say by an amount x.

10. Arthur names a whole number q (chosen so that $(4L/5)^q < x$, where L is Arthur's evaluation of the leftovers). Naming q in advance prevents the next phase from going on forever.

11. Arthur cuts the leftovers into five pieces (a relic of the original idea of Taylor).

12. Bertha trims up to two pieces, if necessary, to create a three-way tie for largest (in her opinion) and sets the trimmings aside.

13. Clare then trims one piece, if necessary, to create a two-way tie for largest (in her opinion).

14. Denis chooses first, then Clare, then Bertha, and Arthur takes the last piece left. If either Clare or Bertha trimmed a piece, they must choose such a piece if it is available at their turn.

15. Steps 11–14 are repeated $q - 1$ further times, each time working with the 'leftovers' from the previous cycle.

    At the end of this cycle of subdivisions, we have an envy-free partial allocation in which Arthur has an irrevocable advantage over Bertha: he thinks his piece is bigger than hers plus all the leftovers. We now create a list of ordered pairs of players, the *irrevocable advantage list*, by writing down the pair (Arthur, Bertha). This reminds us that Arthur has an irrevocable advantage over Bertha.

16. Bertha cuts the leftovers into 12 (equal) pieces.

17. Each of the other three players declares themselves to be 'pro' if they think all these 12 pieces are the same size, or 'con' of not.

18. If every pro has an irrevocable advantage over every con (see the list) then we give the 12 pieces to the pros, each receiving an equal number, and stop.
    (This is why we use 12: it is divisible by 1, 2, 3, and 4.)

19. If not, we choose the first (pro, con) pair where there is no such irrevocable advantage, and return to step 4 with the pro player in the role of Arthur, the con in the role of Bertha, and the leftovers in place of the cake.

20. Repeat steps 5–18 until (after at most 15 cycles) every (pro, con) pair is on the irrevocable advantage list: the cycle then stops at step 18.

For a full proof that this works, see the paper by Brams and Taylor listed under Further Reading.

13

Squaring the Square

The problem of squaring the circle goes back to the ancient Greeks, but that of squaring the square is considerably more recent. Can you tile a square using square tiles? Easy, you might say – consider a chessboard, where 64 small squares tile one big one. However, there's an extra condition: all the tiles must be of different sizes. So how do you solve it? Using electrical circuit theory, naturally!

Can you tile a square using square tiles, all of different sizes? It sounds easy – just experiment. But there are too many possibilities to be sure you've tried them all, and very few arrangements work. Something more systematic is needed. The earliest significant result on this question goes back to 1903, and the answer did not come until 1939. Subsequent work has imposed additional requirements on the tiling, and extended the problem to shapes other than a square. Many open questions remain for the recreational mathematician to investigate.

In 1903 Max Dehn proved that if a rectangle is tiled by squares – whether different or not – then the sizes of the tiles, and of the rectangle itself, are commensurable – integer multiples of a single number. In other words, if we choose a suitable unit of measurement, all sides are whole numbers. This theorem has since been proved in at least a dozen different ways – all of them fairly cunning, because it is by no means an obvious result.

Say that a rectangle or square is *squared* if it can be tiled by distinct square tiles. In 1909 Z. Morón discovered the first squared 33 × 32 rectangle, using nine square tiles of sides 1, 4, 7, 8, 9, 10, 14, 15, and 18. He also managed to tile a 65 × 47 rectangle with ten square tiles of side 3, 5, 6, 11, 17, 19, 22, 23, 24, and 25 (Figure 45a). You might like to make up the nine-square set and try this puzzle: for the answer, see *Unsolved Problems in Geometry* (Further Reading).

The 'squared square' problem was solved in 1939 by R. Sprague, with a tiling that employed 55 different square tiles. However, it lacked

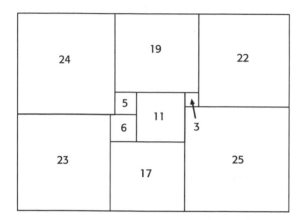

Figure 45

(a) Morón's ten-tile rectangle.

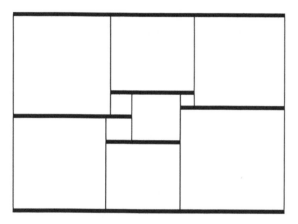

(b) Its horizontal line segments.

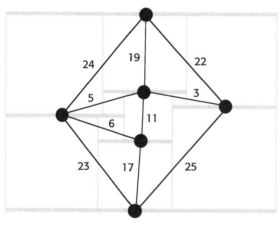

(c) Its Smith Diagram: associate a node with each line segment, and edge with each tile, and label the edge with the size of the tile. Current flows down the page.

elegance in one respect: it was *compound*, meaning that it contained a smaller squared rectangle. Tilings that do not include any squared rectangles are said to be *simple*, and squared squares with this extra property are harder to find.

In 1940 R. L. Brooks, C. A. B. Smith, A. H. Stone, and W. T. Tutte (Further Reading) discovered the first simple squared square. Their method has been entertainingly described by Martin Gardner in *More Mathematical Puzzles and Diversions* (Further Reading). They started by representing the structure of any squared rectangle by a network, known as its Smith Diagram. Every horizontal line in the squared rectangle corresponds to a node of the network, and each component tile corresponds to an edge. This edge links the two nodes corresponding to the horizontal lines that meet the top and bottom of the tile, and the edges are labelled with the size of this tile. Figure 45b shows the horizontal lines in Morón's squared rectangle, and Figure 45c shows its Smith Diagram.

Remarkably, if each edge of the Smith Diagram is assumed to be a wire of unit resistance, and the numerical labels are interpreted as electric currents flowing through the wires (measured in ampères, say, and in the downwards direction on the page) then the whole diagram forms an electrical circuit that obeys the usual 'Kirchhoff's Laws' of electrical engineering. In particular, the total amount of current that flows into any junction must equal the total amount that flows out. This fact follows easily from the geometry of the tiling. For example, consider the horizontal line along the bottom of square 19. A current of 19 flows in, and $5 + 11 + 3 = 19$ flows out. These two numbers are the same because the line also forms the top of the squares of size 5, 11, and 3.

By using ideas from electrical circuit theory, the four mathematicians developed systematic methods for constructing and analysing squared rectangles, with the eventual objective of finding a simple squared square. The first significant breakthrough came from an unexpected quarter. Brooks had found a 112 × 75 squared rectangle with 13 tiles

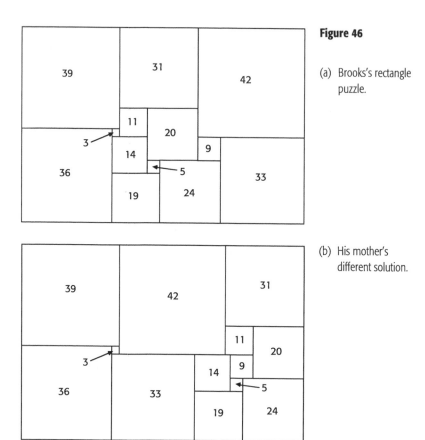

Figure 46

(a) Brooks's rectangle puzzle.

(b) His mother's different solution.

(Figure 46a), and he was so pleased with it that he built a 'jigsaw puzzle' from the tiles. His mother had a go at the puzzle, and succeeded in solving it – but her solution (Figure 46b) was different from Brooks's. The team of mathematicians had never come across such a phenomenon before – a set of square tiles that tiled the same rectangle in two different ways. But they had been hoping for some time to find two squared rectangles of the same size that had no common sizes of tile, because then it would be easy to fit them together, with two extra squares, to make a squared square (Figure 47). It would be a compound square, but it would be a start – and at that stage Sprague had not yet published his discovery.

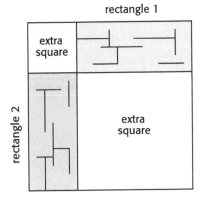

rectangle 1

Figure 47

How to combine two squared rectangles to get a squared square, provided the rectangles have no tiles of common size.

Brooks's rectangle could certainly be tiled in two ways, but since the same set of tiles was used both times, Brooks's rectangle would not lead to a squared square. Nevertheless, they hoped that if they understood *why* Brooks's rectangle could be tiled in two ways, they might get a useful angle on the problem. On looking at the Smith Diagrams of the two tilings they realized that they could obtain one diagram from the other if they 'identified' two nodes – that is, conceptually considered them as being the same. Moreover, the flow of electricity through the circuit was not affected by 'short circuiting' the diagram in this manner, because in this particular instance the two points that were identified were at the same electrical potential. With some effort they worked out why this was the case – it was related to symmetries in the Smith Diagram. From this clue they developed other ways to tinker with Smith Diagrams, producing different squared rectangles of equal size, but having fewer tile sizes in common. Eventually this approach paid off, leading them to a simple squared square formed from 69 tiles. With further effort, Brooks reduced the number of tiles to 39.

In 1948 T. H. Willcocks reduced the number of tiles further, finding a squared square with 24 tiles. But his square was not simple. Meanwhile J. W. Bouwkamp and colleagues were cataloguing all possible squared rectangles with up to 15 tiles – finding a total of 3663. In 1962 A. W. J. Duivestijn proved that any simple squared square must contain at least

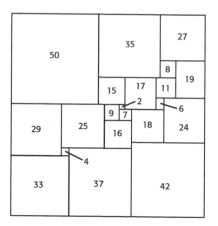

Figure 48

The unique squared square with the fewest tiles.

21 tiles; by 1978 he had found such a square, and proved that it was the only one (Figure 48). In 1992 Bouwkamp and Duivestijn published 207 simple squared squares with between 21 and 25 tiles – all such squares, in fact.

Although these results pretty much polish off the squared squares problem, there are innumerable variants. What about squared dominoes – rectangles with one side twice as long as the other? There is a trivial way to square such a rectangle: start with a squared square and add one additional square tile, the size of the entire squared square, next to it. But are there *non*-trivial ways? What about a rectangle with one side three times as long as the other?

Another extension of the problem is to tile surfaces other than squares and rectangles, an option discussed by David Gale in *The Mathematical Intelligencer* (see Further Reading). Topologists know that several interesting surfaces can be constructed by identifying opposite edges of rectangles – 'gluing them together', at least in the imagination. Take a rectangle and glue two opposite edges: you get a cylinder. Give the rectangle a half twist before gluing, and you get a Möbius band. Glue *both* pairs of opposite edges together, with no twists, and the result is a torus – a doughnut-shaped surface, hole and all. Glue both pairs of opposite edges together, giving one pair a half-twist, and you

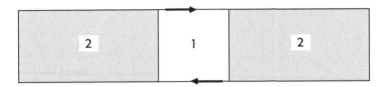

Figure 49 Squaring a Möbius band with two tiles.

get a Klein bottle – a famous one-sided surface that cannot be formed in three-dimensional space without crossing itself. Give both pairs a half-twist, and you get a projective plane – also one-sided and impossible to represent in three dimensions.

Clearly, with any of these ways to glue the rectangle's edges, any tiling of the rectangle carries over to a tiling of the resulting surface. But the surface may possess additional tilings, because tiles on the surface can cut across the glued edges. For example, Figure 49 shows a Möbius band tiled with just *two* squares, of side 1 and 2 respectively. The arrows mark sides to be glued: the directions of the arrows show the twists. Although one square seems in the picture to be cut into two rectangles, those pieces join up when the edges are glued. However, this tiling of the Möbius band has a rather nasty feature: the small tile has a common boundary with itself. Its top and bottom edges get glued – so really it is a Möbius band itself, not a square. In 1993 S. J. Chapman found a tiling of the Möbius band without this awkward feature, using five tiles (Figure 50). No such tiling exists with fewer tiles.

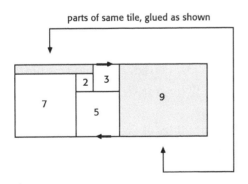

parts of same tile, glued as shown

Figure 50

Squaring a Möbius band using tiles whose boundaries do not meet themselves.

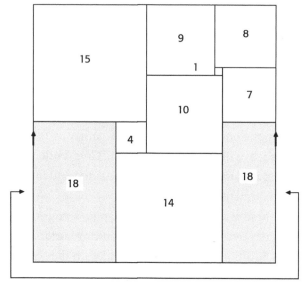

Figure 51
Non-trivial squaring of a cylinder.

parts of same tile, glued as shown

A cylinder can be squared, but this requires at least nine tiles – just as for rectangles. 'Trivial' squarings just take Morón's rectangles and join up suitable edges; but there are also two non-trivial nine-tile squarings. Their tiles are the same sizes as Morón's, but the arrangement is different (Figure 51).

For the cylinder and Möbius band, the edges of the tiles have to be parallel to those of the surface. However the torus, Klein bottle, and projective plane do not have any edges, so tiles could conceivably be set at an angle. In fact, if this is done, a torus can be tiled with just two square tiles (Figure 52) provided we allow two edges of the same tile to meet. (As a bonus, there is a proof of Pythagoras's Theorem hidden in this picture: can you see why?)

Not much seems to be known about tilings of the Klein bottle. Every tiling of a Möbius band can be glued along its edge (there is only one) to give a tiling of the Klein bottle, and there are no other ways to tile the

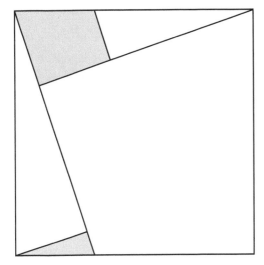

Figure 52

Squaring a torus with just two tiles – and a hidden proof of Pythagoras. (Hint: look at the right triangle whose hypotenuse is the left-hand edge of the diagram.)

Klein bottle with six or fewer square tiles. Nobody knows whether this remains true for seven or eight tiles – but it is false for nine.

Virtually nothing is known about tilings of the projective plane. And what about tiling, say, the surface of a cube? The whole field is wide open.

14

The Bellows Conjecture

Can a polyhedron with triangular faces be flexible? Contrary to conventional wisdom, it turns out that some of them can. Now it has been proved that if a polyhedron can flex then its volume cannot change, because of a remarkable formula that the classical mathematicians missed. So why does a concertina work, then?

Every amateur carpenter who has tried to build a bookcase knows that rectangles are not rigid. If you lean against the corner of a rectangle then it tilts sideways to form a parallelogram (Figure 53a) – and in all likelihood collapses completely. A triangle, on the other hand, *is* rigid: it cannot be deformed without changing the length of at least one side. Euclid knew this, in the form 'if two triangles have sides with the same lengths, then the triangles are congruent (have the same shape)'. In fact, the triangle is the *only* rigid polygon in the plane. Any other polygonal shape must be braced in some manner. For example, cross-struts can be added, to break it into triangles (Figure 53b), or shapes that are themselves rigid can be assembled in threes (Figure 53c).

Another way to rigidify your bookcase is to nail a flat back onto it. This takes the question into the third dimension, where everything becomes far more interesting, and surprises abound. For nearly 200 years mathematicians have been puzzled by the rigidity, or otherwise, of polyhedrons – solids with finitely many faces, which are polygons that meet in pairs along edges. Until recently it was assumed that any polyhedron with triangular faces must be rigid – but that turned out not to be true. There exist 'flexible' polyhedrons, which change shape even though no face distorts or bends by even the tiniest amount – I'll come back to those in a moment.

The latest discovery, made by Robert Connelly (Cornell), Idzhad Sabitov (Moscow State University), and Anke Walz (Cornell), is that flexible polyhedrons cannot change their volume. It is not possible to make a polyhedral 'bellows' that can flex and blow air out through a

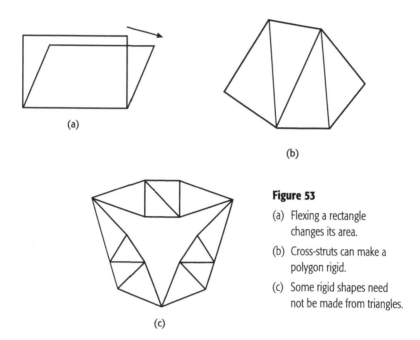

(a)

(b)

(c)

Figure 53

(a) Flexing a rectangle changes its area.

(b) Cross-struts can make a polygon rigid.

(c) Some rigid shapes need not be made from triangles.

hole as its internal volume shrinks. (What about concertinas? See below.) Their proof required them to discover some unexpected properties of polyhedrons, which are likely to prove important in future research.

Before starting on the mathematics, I'd better make one thing clear. Anyone who has folded origami figures from paper knows that it is possible to make birds that flap their wings, frogs whose legs move, and so on. Aren't these flexible polyhedrons? The answer is 'no', for two reasons. One reason is that the paper has edges, so it does not form a polyhedron. The other, more important, reason is that when the paper frog moves its legs, the paper *bends* slightly. The same goes for concertinas, which at first sight appear to be polyhedral bellows, but again these work only because of slight bending (and perhaps even a little stretching). From now on, no amount of bending, not even by a trillionth of a micron, will be permitted. When a polyhedron flexes, the *only* things that can change are the angles at which faces meet. Imagine that the faces are hinged along their edges, and flex the hinges. All else is perfectly rigid.

The whole area dates back to 1813, when the great French mathematician Augustin Louis Cauchy proved that a convex polyhedron – one without indentations – cannot flex. But what if there are indentations? The first flexible non-convex polyhedron was found by Raoul Bricard, a French engineer – except that in his example faces were permitted to interpenetrate freely, and move through each other. This is of course impossible for a real physical object. However, Bricard's example can be realized if we remove the faces and replace the edges by rigid rods to get a *linkage*. Bricard also invented chains of simple polyhedrons, joined edge to edge, that can flex. According to W. W. Rouse Ball's famous book *Mathematical Recreations and Essays* (Further Reading), the simplest such rings were invented by J. M. Andreas and R. M. Stalker. These are rings of six or more regular tetrahedrons – the number must always be even – hinged together along pairs of opposite edges (Figure 54). With six tetrahedrons the amount of movement is slight, but with eight or more, the ring can rotate indefinitely, like a smoke ring. With 22 or more, the ring can even be knotted! However, such shapes are not true polyhedrons because more than two faces meet along some edges.

The topic did not really come alive until the 1970s, when Connelly

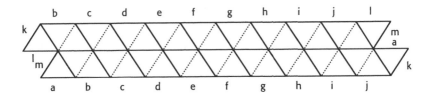

Figure 54 A ring of ten tetrahedrons. Fold solid lines into ridges, dotted ones into valleys. Join tabs with the same letter.

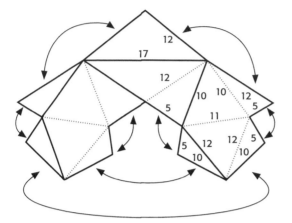

Figure 55

Steffen's flexible polyhedron. Fold solid lines into ridges, dotted ones into valleys.

modified Bricard's self-penetrating flexible polyhedron in such a manner that it remained flexible, but ceased to be self-penetrating. Within a few years the construction had been simplified by Klaus Steffen (University of Düsseldorf), to yield a flexible polyhedron with nine vertices and 14 triangular faces (Figure 55). It is amusing to make a model out of thin card, and see how it flexes. As far as anyone knows, this is the simplest possible flexible polyhedron, but it is very difficult to see how to go about proving such a statement.

Mathematicians who investigated these and other flexible polyhedrons quickly noticed that as they flexed, some parts moved closer together while others moved further apart. Qualitatively, at least, it looked as if the total volume might not change during the motion. Dennis Sullivan (City University of New York) filled a flexible polyhedron with smoke, flexed it, and observed that no smoke puffed out. This elegant but crude experiment suggested – but of course did not prove – that the volume remained unchanged. And so the Bellows Conjecture was born. It states that a flexible polyhedron has constant volume while it flexes – a polyhedral bellows is impossible.

The first interesting feature of the Bellows Conjecture is that its planar analogue is false. When a flexible polygon, such as a rectangle, collapses into a parallelogram, the area gets smaller. Clearly there is something unusual about three-dimensional space that makes a bellows impossible. But what? Connelly's group focused on a famous formula for the area of a triangle, believed to be due to Archimedes, but usually credited to Heron of Alexandria who wrote down a proof. Heron was a Greek mathematician who lived somewhere between 100 BC and AD 100, and he stated and proved the formula in his books *Dioptra* and *Metrica*. The formula is shown in the box, but what matters here is not so much the details, as the general nature of the formula. It can be rearranged, using algebra, to give an equation relating the area of the triangle to its three sides. Moreover, this equation is polynomial: its terms are just whole number powers of the variables, multiplied by fixed numbers.

HERON'S FORMULA

Suppose that a triangle has sides a, b, c, and area x. Let s be the semi-perimeter:

$$s = (a + b + c)/2$$

Then

$$x = \sqrt{s(s - a)(s - b)(s - c)}$$

Square this equation and rearrange to get rid of the 1/2's: the result is

$$16x^2 + a^4 + b^4 + c^4 - 2a^2b^2 - 2a^2c^2 - 2b^2c^2 = 0$$

This is a polynomial equation relating the area x to the three sides a, b, c.

Sabitov came up with the curious – and at first implausible – idea that there might be a similar polynomial equation for *any* polyhedron, relating the polyhedron's volume to the lengths of its sides. Such a polynomial would be a truly remarkable discovery, because until that moment nobody had suspected that any such thing could possibly exist. Yes, there were some well-known special formulas – easy ones for cubes and rectangular boxes, and something a bit like Heron's formula for tetrahedrons (solids with four triangular faces, regular or irregular pyramids on a triangular base) only more messy. But nothing completely general, applying to any polyhedron.

Could the great mathematicians of the past really have missed such a wonderful idea? It seems unlikely.

Nevertheless, suppose such a formula *does* exist. Then the Bellows Conjecture is a simple consequence. The reason is straightforward. The formula relates the volume to the sides. As the polyhedron flexes, the lengths of its sides don't change – so the formula stays exactly the same. Its solution, the volume, must therefore also stay the same.

Actually, there is one technical point to take care of. A polynomial equation can have *several* distinct solutions, so in principle the volume might suddenly jump from one solution to a different one. However, the volume obviously changes *gradually* if the flexing is gradual, so whatever the volume does, it cannot jump. End of proof.

When I first wrote about this, some readers told me that it must be wrong. For example, if you make a box-shaped house with a roof and turn the roof upside down, all the sides have the same length as before but the volume gets much smaller (Figure 56). That's true, but it doesn't invalidate the reasoning. First, the same lengths can correspond to several different polyhedrons (as they do here) because, as just pointed out, a polynomial equation generally has several solutions. The number must be finite, but the larger the degree of the polynomial, the more solutions there can be. Second: you can't deform the normal house into the other one in a continuous manner without faces getting bent and buckled along the way. Objection overruled!

At any rate, the problem will be solved if we can find a polynomial equation for the volume of a polyhedron in terms of its sides. There is an obvious place to start: the classical formula for the volume of a tetrahedron – the simplest polyhedron. Just as any polygon can be divided into triangles, so can any polyhedron be divided into tetrahedrons. Then the volume of the polyhedron is just the sum of the volumes of

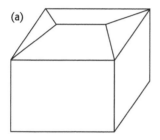 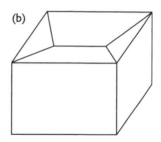

Figure 56

(a) House with roof.

(b) House with reversed roof. Both have the same edges but different volumes.

those tetrahedral pieces. But that won't, of itself, solve the problem. The formula that it leads to involves all the edges of all the pieces, but many of those are not edges of the original polyhedron. Instead, they are various 'diagonal' lines that cut across from one corner of the polyhedron to another one, whose lengths may very well change if the polyhedron flexes. So somehow the formula has to be massaged, algebraically, to get rid of those unwanted edges and 'glue' all the component equations together into one Grand Unified Equation.

It was always going to be messy. For an octahedron, with eight triangular faces, it turned out that such a massaging procedure was possible, but the resulting equation involved the 16th power of the volume. More complex polyhedrons would surely require higher powers still. However, the octahedron was a good start. By 1996, Sabitov could write down an explicit but extremely complicated procedure for finding suitable equations. In 1997 the team of Connelly, Sabitov, and Walz found a far simpler way to achieve the same result.

The reasons why such equations exist are not fully understood. In two dimensions, they don't – except for the rigid triangle and the Heron equation. In three dimensions, we now know that they do. Connelly and Walz think they know how to prove a four-dimensional

Bellows Conjecture. For five dimensions or more, the problem is wide open. But it's fascinating to see how a simple experiment with some bits of card and some smoke opened up a marvellous, totally unexpected, and fundamental mathematical discovery. And there *are* simple ideas that the great mathematicians of the past could have discovered – but didn't.

15

Purposefully Piling Pyramids

How many men does it take to build a pyramid? More than it takes to change a light bulb, presumably – although one man could do it if he lived long enough. A million would do the job faster, if they didn't trip over each other. The correct answer must be somewhere in between – but roughly how big is it? The Roman historian Herodotus claimed that the Great Pyramid was built using 100,000 slaves. Archaeology has shown that the workers weren't slaves. Mathematics shows that 100,000 is far too big. It's all a matter of energy.

The pyramids of ancient Egypt rank among the most enigmatic of archaeological mysteries. Several of them are enormous. Khufu's 'Great' pyramid at Giza, which dates from about 2500 BC, originally had a volume of over 2.5 million cubic metres and a mass of 7 billion kilograms. It is made from huge stone blocks, which were quarried, trimmed to a fairly regular shape, transported to the construction site, and then piled on top of each other with astonishing precision.

How did the ancient Egyptians put together such vast edifices? What were they for? How were they built? Mostly, we don't know – although we do know that many pyramids were used as tombs for kings, and there are plenty of theories about the pyramids' construction. However, thanks to some clever mathematical detective work by Stuart Kirkland Wier of Denver Museum of Natural History (Further Reading), we now have a good idea what size the workforce was.

Some two millennia after the pyramids were constructed, the Roman historian Herodotus reported that it took 100,000 men to build the Great Pyramid. However, Herodotus is not a reliable source, and it now seems that he overestimated the workforce by an order of magnitude. According to Weir, the true figure was probably around 10,000 – surprisingly small. How can we be sure about the number of workers when we have no clear idea of how the pyramids were built? Provided we make a few reasonable assumptions about how the Egyptians went about their business, and granted that they weren't totally incompetent, we can estimate the workforce from a few simple mathematical principles.

For definiteness, Wier works with Khufu's pyramid, but the same method can be applied to other pyramids, with roughly similar results. The main idea is to work out how much energy a pyramid contains. By 'energy' here I mean 'potential energy' – the effort required to lift a given mass through a given height against the force of gravity. If we divide the potential energy of the pyramid by the number of days needed to build it, we get the mean energy required per day in order to lift the blocks. All we now need is an estimate of how much energy per day a typical Egyptian construction worker would be able to provide, and we can work out the average size of the workforce. Just divide the total energy per day by the energy expended in a typical man-day.

As it stands, such a calculation makes several assumptions. It omits any other activities that require energy, such as transporting the materials, cutting the stone, building machinery – even feeding the workforce. It places a lower limit on the number of workers, but not an upper limit. And even if we know the average size of the workforce, we can't be sure how the actual size varied around that average. In order to improve the estimate, we must determine approximate values for any other important energy requirements, consider how efficient the actual energy usage would have been, and try to figure out the likely pattern of construction. Did the pyramid-builders employ a workforce of fixed size? Or did they hire extra workers when the project demanded it and fire them when they were no longer needed? The records don't tell us, but we can infer rather more than might be expected by assuming that the Egyptians behaved like sane, reasonable people.

The biggest uncertainty is time. How long did it take to build Khufu's pyramid? Khufu reigned for 23 years. Construction of his pyramid would not have begun much before his reign did, and it would have ended either before his death or soon afterwards. On the other hand, the construction boss would not have known for sure when Khufu would die. So, for a ball-park estimate, the simplest assumption is that Khufu's pyramid took 23 years to build – the same length as the king's reign. That amounts to 8400 days, assuming building continued all the year round. The actual time might have been half as long, or twice as long: because of this fundamental uncertainty about timing it is pointless to enter into fine detail when estimating other relevant figures.

A certain amount about pyramid building techniques can be deduced from ancient records and from the layout of the pyramid site (Figure 57). Pyramids were made from blocks of stone, taken from nearby quarries whose location is sometimes known. The sole source of power was human muscles (no water power, for example) aided by primitive but effective tools such as levers. On the whole, a pyramid was built in layers from bottom to top – certainly no block could be put in place until those below had been. Horizontal transport of the blocks was done by workers dragging wooden sledges (Egyptian carvings

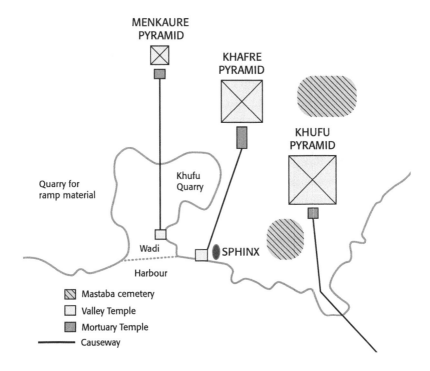

Figure 57 Schematic map of Khufu's pyramid and some of its surroundings.

show this being done). Nobody knows how the vertical transport was achieved: theories include vast ramps of sand, cunning arrangements of levers, and piles of wooden supports.

Khufu's pyramid, when new, was about 146.7 metres high, with a base 230.4 metres square. The volume of a pyramid of height h and base a square of side s is $s^2h/3$, which here works out as 2.59×10^6 cubic metres. The material is limestone, of density $d = 2.7 \times 10^3$ kilograms per cubic metre, so the mass is 7.01×10^9 kilograms. The potential energy of a pyramid (an interesting exercise in calculus) is $gds^2h^2/12$, where g is the acceleration due to gravity (9.81 metres per second). This amounts to 2.52×10^{12} joules, a joule being the relevant unit of energy.

According to *The Engineer's Manual* for 1917, the average amount of useful work provided by a man in a day is about 2.4×10^5 joules. So the number of men that would be needed just to raise the blocks into position, assuming perfect efficiency, is $2.52 \times 10^{12}/(8400 \times 2.4 \times 10^5)$, which is 1253. This estimate is clearly too low, because in practice efficiency is not perfect, but it puts the effort required into perspective – it's not as much as you might expect.

Getting a better estimate requires more thought about logistics. Pyramids are not just featureless heaps of stone: they have passages and chambers, some of which are remarkable feats of engineering in their own right. But the overwhelming part of the work was piling up those stones, so we can ignore structural details. Dividing the volume of the pyramid by the time available to build it shows that about 310 cubic metres of stone per day must be put in place, on average. As the height that a block must be raised increases, the energy needed per block increases too; moreover, the amount of workspace at the top of the pyramid decreases as the pyramid gets higher. These factors make it clear that a steady rate of 310 cubic metres per day is not sensible. Instead, the rate of installation of stone should be greater when the pyramid is low, and drop off as it gains in height.

Wier considers three representative construction schedules for the pyramid:

(A) A constant rate, subject to the practical requirement that the area available to install stone never falls below 10 square metres for every cubic metre installed in a given day.

(B) A construction rate that declines linearly according to the height of the currently built portion of the pyramid.

(C) A construction rate that falls off slowly, then more rapidly, and finally tails off again.

These schedules are not proposed as models of what the Egyptians actually did: they are representative possibilities to be used as guides.

Assuming 8400 days for completing each schedule, Figure 58 shows how the construction rate varies with the height of the currently built portion, and Figure 59 shows how the construction rate varies with time. For example, schedule (A) requires 315 cubic metres of stone to be installed every day until 8110 days, after which construction falls off rapidly because of the limited space at the top of the pyramid.

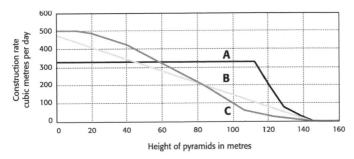

Figure 58 How construction rate varies with height, for three representative schedules.

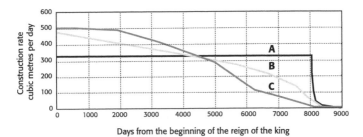

Figure 59 How construction rate varies with time, for three representative schedules.

Once the construction rate is known, the manpower required can be estimated. For instance, take case (B). Here the construction rate starts at 462 cubic metres per day when the pyramid has height 0 metres, and declines by roughly 31 cubic metres per day for every 10 metres of extra height. A useful mathematical trick is to split the movement of stones into two components: vertical ('lifting') and horizontal ('hauling'). We are not here assuming that these two components actually *happen* separately – for example, if the stones are dragged up a sloping ramp then both components change simultaneously. It's just easier to calculate the two contributions separately.

The vertical part includes not just lifting the stones up the pyramid, but raising them from the bottom of the quarry to the level of the base of the pyramid – a distance known to be 19 metres. The number of men required to lift the stones can be calculated from the potential energy needed, multiplied by a fudge factor to account for inefficient use of muscle-power. The horizontal part includes sliding the stones from the quarry to the pyramid – a distance of 635 metres or less – and any further horizontal movement needed to set them in place on the pyramid itself. In order to find the number of men required for hauling, we estimate the friction between a wooden sledge and the sandy ground, and calculate the work done against friction, which gives the energy involved.

Manpower is also required for other tasks – quarrying the stones, trimming them to shape, making wooden sledges, whatever. Wier assumes that all such tasks combined require between 5 and 10 men per cubic metre of stone per day. The results, for case (B), are shown in Figure 60, taking the upper estimate of 10 men per day. At no stage does the workforce exceed 12,800 men – slightly less than 1% of the Egyptian workforce at that time. Cases (A) and (C) lead to very similar results, making it clear that any reasonable construction schedule would involve a workforce of roughly that size.

Perhaps the simplest schedule is to employ a workforce of constant size, except near the end where there isn't enough room at the top for

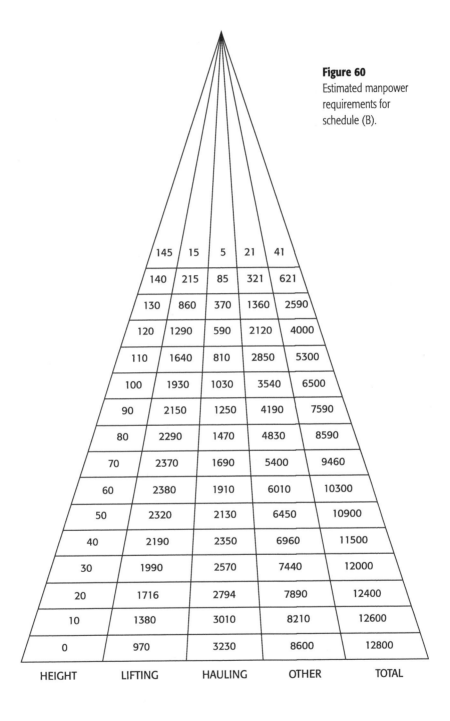

Figure 60
Estimated manpower requirements for schedule (B).

HEIGHT	LIFTING	HAULING	OTHER	TOTAL
145	15	5	21	41
140	215	85	321	621
130	860	370	1360	2590
120	1290	590	2120	4000
110	1640	810	2850	5300
100	1930	1030	3540	6500
90	2150	1250	4190	7590
80	2290	1470	4830	8590
70	2370	1690	5400	9460
60	2380	1910	6010	10300
50	2320	2130	6450	10900
40	2190	2350	6960	11500
30	1990	2570	7440	12000
20	1716	2794	7890	12400
10	1380	3010	8210	12600
0	970	3230	8600	12800

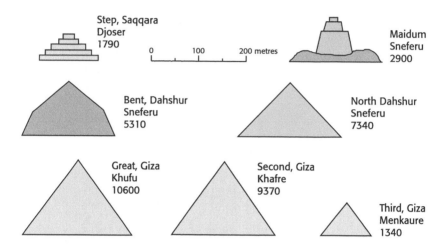

Figure 61 Estimated workforce for a variety of surviving pyramids.

more than a few workers. It turns out that, with the same assumptions, 8300 men then suffice to build Khufu's pyramid. Similar results hold for other pyramids (Figure 61). Despite the astonishing size of the pyramids, it seems that manpower was not a problem.

How did the Egyptians lift so many stones into place? A common theory is that they used huge ramps made of sand, which were later removed. They certainly used small ramps for various construction purposes, but it seems unlikely that they built gigantic ramps rising to the top of the pyramid. It takes a lot of energy to build a ramp – roughly as much as it does to build a complete pyramid – and then the ramp has to be removed again. Another problem with ramps is that the workers don't just have to lift the stones against gravity: they have to lift *themselves* as they make their way up the ramp. Unless they haul on ropes, which isn't easy if the ramp is a long one.

A better way would be to keep the movement of workmen as low as possible. One simple and efficient way to achieve this was suggested to me a few years ago by Alan Moore, an Englishman who has long been interested in puzzles about ancient technology. His idea is to use a series of levers running up the face of the pyramid. Stones secured in rope 'bags' could be rapidly passed from lever to lever, moving up the pyramid one step at a time. The men working the levers would not need to move up or down themselves, except when changing shift. As soon as a lever has raised a stone, it is ready to raise the next one. Apart from the initial effort of building the levers, which is tiny in comparison to the rest of the work, this method wastes very little energy, and – a major plus – needs very little workspace.

We may never know just what the Egyptians did, but it's instructive to look for practical methods and work out how much effort they would require. And some of the results, as Wier realized, are largely independent of the methods used, because of simple, universal mathematical principles.

16

Be a Dots-and-Boxes Grandmaster

We all played the game when we were kids. Draw a rectangular grid of dots, and take it in turns to join neighbouring ones. If you enclose a square box, initial it: it's yours. And you must play again, until you stop completing boxes. Whoever owns the most boxes wins. Simple? The rules are; the strategic implications are not. Dots-and-Boxes (or plain Boxes) is one of the most fiendishly complicated games ever invented. And very few people even get to the first level of competent play – or realize that it exists.

I **never cease to be amazed by the mathematical subtleties** inherent in what seem to be the simplest of games. Even children's games raise questions that require sophisticated mathematics. Elwyn Berlekamp's book *The Dots and Boxes Game* (Further Reading) takes this particular line of thought to new heights. Virtually everyone has played this game in primary (grade) school – but having read the book, I doubt whether one person in a million has played it anywhere near the highest standard attainable. Few players are aware of even the first layer of subtlety in the strategy for the game. Berlekamp makes it clear that there are plenty of deeper layers – and unplumbed depths remain.

Let's remind ourselves of the rules. Play begins with a rectangular grid of dots. Players take turns to draw a line joining two dots that are adjacent either horizontally or vertically (but not diagonally). If a player completes the fourth side of a box – a square of unit side – then they write their initial in that box and play again (and *must* continue to play again, as long as they keep forming completed boxes). At the end of the game, the player who has initialled the most boxes wins.

Let's call the players Alfred and Betsy, and make the standard convention that Alfred always starts. Figure 62 shows typical play by children (and most adults): I'll call this Level 0 play. Players avoid giving away boxes for as long as they can, by finding a move that does not create the third side of any potential box. As a result, the grid becomes divided into a series of 'chains'. These are snake-like regions bounded by lines, such that as soon as a player claims one box at the

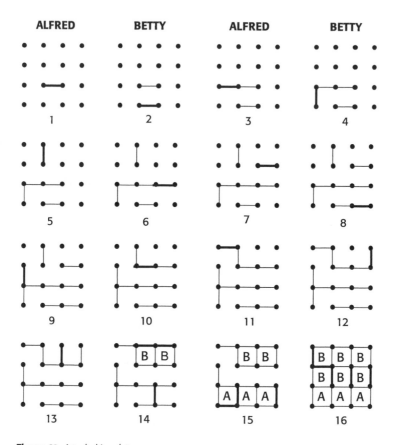

Figure 62 A typical Level 0 game.

start of the chain, they can continue grabbing boxes until the entire chain is used up. Chains can close up into loops.

At some point, the grid has been fully divided into such chains – a state that I'll call *gridlock*. Here such a state is created by Betsy on the 12th move. After gridlock is reached, players usually draw the next line in the shortest chain available, thereby giving the opponent the smallest number of boxes. Having completed this chain, the opponent then gives away the next smallest chain ... and so it goes.

In the game illustrated, there are three chains at gridlock: they contain 2, 3, and 4 potential boxes respectively. With Level 0 play, Alfred is eventually forced to give away a chain of length 2 to Betsy. Betsy in turn cedes a chain of length 3 to Alfred, and Alfred is then obliged to present Betsy with a chain of length 4. Betsy wins by 6 boxes to Alfred's 3.

Level 1 play improves on Level 0 by keeping track of who wins once gridlock is attained, and trying to choose moves to make sure it's you and not your opponent. In Level 0 play, the winner depends on the parity (oddness or evenness) of the number of chains when gridlock is reached, *and* on which player cedes possession of the first chain to their opponent by playing inside that chain immediately after gridlock. Let's say they 'open' the chain: note that you open a chain by presenting it to your opponent – you don't win it yourself.

If the number of chains at gridlock is even, then the player who opens the first chain wins, because every chain their opponent completes is equalled or beaten by their own next chain. Note that in this situation the player who opens the first chain makes the last move of the game. On the other hand, if the number of chains is odd, then the player who opens the first chain loses – because their opponent gets the first chain, and every subsequent chain that they get is equalled or beaten by their opponent's next chain. In this situation the opponent makes the last move of the game.

In our example, there are three chains (odd) at gridlock, and the number of moves taken to reach gridlock is 12 (even). Alfred is forced to open a chain, so Betsy gets the first and third chains; Alfred has to make do with just the second.

Which player is the first to break gridlock depends on the parity of the number of moves made to reach that state. If this number is even, then Alfred opens up the first chain and Betsy makes the first territorial gain; if it is odd, Betsy opens up the first chain and Alfred makes the first territorial gain. If Alfred wants to win a game where Betsy is playing Level 0, he has to ensure that the number of moves made to reach gridlock, *plus* the number of chains that result, is even. If so, then either Alfred opens up the first of an even number of chains, or Betsy opens up the first of an odd number of chains. Either way, Alfred wins.

However, Betsy can play Level 1 too. If Alfred is playing Level 1, then she has to ensure that the number of moves made to reach gridlock, *plus* the number of chains that result, is odd. Careful consideration of alternatives, a few moves away from gridlock, can help her achieve this goal, but won't always guarantee it.

Even at this low level of play, we see that there are simple mathematical principles, to do with the parities of various numbers associated with the state of play. However, Level 2 play knocks these particular principles on the head, by refusing to play the Level 1 strategy if it will lead to a loss. In this example, Alfred knows that if both players use the Level 1 strategy after position 12, then he will lose. He therefore comes up with a cunning plan to put Betsy at a disadvantage. At move 13 he still opens up the chain of length 2, presenting Betsy with two boxes. Her move 14 opens up the chain of length 3 for Alfred to take. But on move 15, Alfred declines to accept all three boxes in that chain. Instead, he accepts one of them, and then draws a line that leaves an enclosed 2 × 1 rectangle, which I'll call a *domino* (Figure 63).

This is known as a *double-dealing* move. It is a 'sacrifice' – it offers Betsy an easy territorial conquest – but it puts her in a fatal position, whether she accepts the sacrifice or not. If she plays in the domino, she

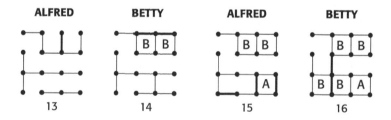

Figure 63 A Level 2 improvement.

gains two boxes *but* she has to play again, opening the chain of length 4, and Alfred swipes the lot, winning by 5 boxes to 4. If instead she opens up the chain of length 4, however, then it's even worse: Alfred takes the whole chain and then takes the two boxes in the domino as well, winning by 7 boxes to 2.

In this case, Betsy's cause is clearly lost the moment that Alfred makes his double-dealing move, because there is only one chain left, plus the domino. Suppose, however, that there are several chains left. Can't Betsy claw back some territory by making double-dealing moves herself?

Not always. Suppose a position is reached where there are a number of dominos, together with some chains of length 3 or more, which we'll call 'long chains' from now on. Suppose Betsy has to play. She may as well grab all available dominos – if not, Alfred can grab them in the middle of his next series of moves without putting himself into a worse position. So now she tries to play a modified Level 1 strategy: open up the shortest remaining long chain. Then Alfred will be forced to open up the next shortest ... so everything depends on the parity of the number of long chains, much as before. Yes?

No. If the number of long chains left is odd, then certainly Alfred would be happy to play a Level 1 game. But if it is even (indeed if it is odd) he is not obliged to accept the whole chain. Instead, he can play Level 3 strategy, taking *all but two* of the boxes, and close with a double-dealing move. Now Betsy has the same problem as before, but with one long chain gone (implying a change of parity). In fact, Alfred can politely hand two boxes per chain to Betsy, keeping the rest for himself, and force her to keep opening up new long chains.

If these long chains contain 5 or more boxes, Alfred wins every time. If they contain 4, he and Betsy split them. Only if they contain 3 boxes does this strategy yield Betsy one box more than Alfred. If there are several long chains, and enough have length 5 or more to cancel out the losses incurred on those of length 3, then Alfred wins.

Say that a player *has control* of the game if they can force their opponent to open a long chain. Then what we've understood so far can be summarized as follows. A good way to play (maybe not the best way, but effective against most players) is to gain control, and then retain it by declining the last two boxes of every long chain. Except the last one, of course, when you do better by grabbing the last two boxes as well. If there are several long chains, which is common, then this strategy usually works very well.

In short, the game in not about boxes as such: it is about gaining control. Now we are ready to move on to Level 4 strategy. How do you get control? It turns out that parity once more holds the key, but first we need another concept. When a player accepts a domino by filling in the central stroke, thereby acquiring two boxes with one move, we call this a *double cross*. Then an effective strategy for gaining control is:

- Alfred tries to make the number of initial dots plus the number of double crosses *odd*.
- Betsy tries to make the number of initial dots plus the number of double crosses *even*.

We can state this rule more simply by noting that whatever size the grid is, the number of dots plus the number of double crosses is equal to the total number of moves in the game. A little thought leads to:

- Alfred tries to make the number of initial dots plus the number of eventual long chains *even*.
- Betsy tries to make the number of initial dots plus the number of eventual long chains *odd*.

You may think this is getting pretty deep, but so far we've only reached page 7 out of 86 pages of strategy in Berlekamp's book. Further topics include a closely related game, called Nimstring, and the concept of nim-addition, which is fundamental to many games but would take another chapter, if not ten, to describe adequately. With these techniques under your belt, you will win more frequently in games where the final score is very close.

I will describe one more concept, though: a *loony move*. This is one of the following:

- Completing a chain of length 2 in such a way that the opponent can create a domino (known as a *half hearted-handout*). See Figure 64.
- Opening a long chain.
- Opening a loop of length 4 or more.

Figure 64 Half-hearted handouts.

It can be proved that if your opponent makes a loony move, then you can secure at least half the remaining boxes. However, the proof is not constructive – that is, it doesn't specify how you should play to achieve that end. The idea is that in each case you have a choice between two approaches, and if one of those is good for your opponent then the other is good for you. Nimstring helps to shed some light on which moves are good here.

Most games played between experts eventually reach a position where all available moves are loony. This is the *loony endgame*, and mathematically it gets very complex: Berlekamp analyses it in depth. Winning a loony endgame is mostly a matter of gaining control – so, once more, we're back to control as the key.

Good old Dots-and-Boxes, then, is a much more sophisticated game than most of us imagine – so sophisticated that no complete winning strategy is known. Berlekamp describes it as 'the mathematically richest popular child's game in the world, by a substantial margin.' Never underestimate the mathematical content of simple-looking games.

17

Choosily Chomping Chocolate

Yucky Choccy and Chomp are chocaholic games with very similar rules, but there the resemblance ends. The first is a 'dream game' with an easy winning strategy. The second is a 'nightmare game' — we know that the first player should always win, if they make the best moves, but we don't know what those moves are. Between them, these games teach us a lot about winning strategies and how to find them. Or not.

Just because a game has simple rules, that doesn't imply that there must be a simple strategy for winning it. Sometimes there is – noughts-and-crosses (tic-tac-toe) is a good example. But sometimes there isn't – another childhood game, Boxes, in which players take turns to fill in edges on a grid of dots and capture any square they complete, is a case in point (but see Chapter 16). I call the first kind 'dream games' and the others 'nightmare games', for fairly obvious reasons. Games with very similar rules can be surprisingly different when it comes to their dream or nightmare status. And, of course, the nightmare games are often the most interesting, because you can play them without knowing in advance who ought to win – or, in some cases, knowing who ought to win but not knowing how they can do it.

As an illustration of these surprising facts, I'm going to discuss two games based around chocolate bars. One, 'yucky choccy', is a dream game. The other, 'chomp', has very similar rules, but it's a nightmare game – with the startling extra ingredient that with optimal play the first player should always win, but nobody knows how.

I have no idea who invented yucky choccy: it was explained to me by Keith Austin, a British mathematician at Sheffield University. It takes place on an idealized chocolate bar, a rectangle divided into smaller squares. Players – I'll name them 'Wun' and 'Too' after the order in which they play – take turns to break off a lump of chocolate, which they must eat. Call this action a *move* in the game. The break must be a single straight line cutting all the way across the rectangle along the lines between the squares. The square in one corner contains

Figure 65

Game tree for 4 × 4 yucky choccy. Arrows indicate legal moves: the piece removed is eaten. The square shown in black is the soapy one. Solid arrows indicate an actual game, shaded arrows alternative moves that could have been made instead.

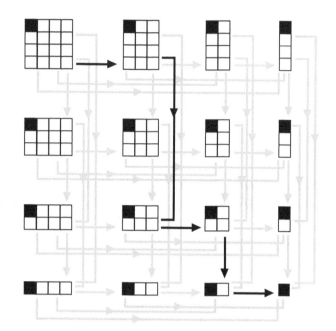

a lump of soap, and the player who has to eat this square loses. The solid arrows in Figure 65 shows the moves in the game played with a 4 × 4 bar, and the shaded arrows show all the other moves that could have been made instead. This entire diagram constitutes the *game tree* for 4 × 4 yucky choccy. As we'll shortly see, Too made a bad mistake and lost a game that should have been won.

A *winning strategy* is a sequence of moves that forces a win, no matter what moves the opponent makes. The concept of a strategy involves not just one game, but all possible games. When you play chess, most of your planning centres on 'what if' questions. 'If I advance my pawn, what could his queen do then?' Tactics and strategy centre around what moves you or your opponent *could* make in future, not just the moves that they *do* make.

There is a neat theory of strategies for 'finite' games – ones that can't continue forever and in which draws are impossible. It relies on two simple principles:

(1) A position is a winning one if you can make *some* move that places your opponent in a losing position.

(2) A position is a losing one if *every* move that you can make places your opponent in a winning position.

The logic here may seem circular, but it's not: it's recursive. The difference is that with recursive reasoning you have a place to start. To see how, I'll use the above two principles to find a winning strategy for 4 × 4 yucky choccy. The trick is to start from the end and work backwards, a process called 'pruning the game tree'.

The single soapy piece ■ is a losing position. I'll symbolize that fact by the diagram

```
L   *   *   *
*   *   *   *
*   *   *   *
*   *   *   *
```

whose entries refer not to a chocolate bar, but to the various positions marked in Figure 65. Here 'L' means 'losing position', * means 'don't know yet', and 'W' will mean 'winning position' once I've found some. In fact, ■☐ ■☐☐ ■☐☐☐ are all winning positions, because you can break off all the white squares in one move to leave your opponent with the single-piece losing position. Equivalently, there are arrows in the game tree that lead from those positions directly to ■, and by principle (1) all such positions are winners. For similar reasons the same positions rotated through a right angle are also winners, so now we have pruned away all branches of the game tree that lead in one step to the single soapy square, which tells us the status of those positions:

```
L   W   W   W
W   *   *   *
W   *   *   *
W   *   *   *
```

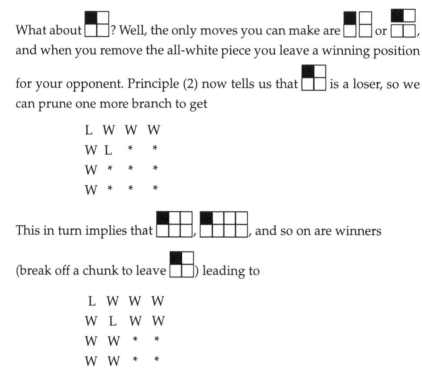

What about 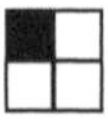? Well, the only moves you can make are  or 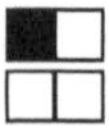, and when you remove the all-white piece you leave a winning position for your opponent. Principle (2) now tells us that 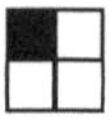 is a loser, so we can prune one more branch to get

L	W	W	W
W	L	*	*
W	*	*	*
W	*	*	*

This in turn implies that 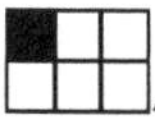, 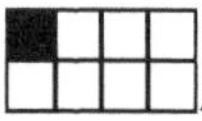, and so on are winners (break off a chunk to leave 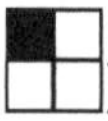) leading to

L	W	W	W
W	L	W	W
W	W	*	*
W	W	*	*

Working backwards in this manner you can eventually deduce the win/lose status of any position. The logic runs not in circles, but in interlocking spirals, climbing down the game tree from leaf to twig, from twig to branch, from branch to limb ... Hence the 'pruning' image. We have to start from the *end*, though, which is a nuisance. What we really want to do, though, is chop down the entire game tree in one blow, George Washington fashion, to find the status of the opening position – and if it's a winner, to find what move to play. For games with a small tree there's no difficulty: repeated pruning yields the status of all positions. In Figure 65 we can carry this out, to get

L	W	W	W
W	L	W	W
W	W	L	W
W	W	W	L

So the 4 × 4 position, for instance, is a loser.

If you try larger bars of chocolate, square or rectangular, you'll quickly find that the same pattern emerges: losers live along the diagonal line, all other bars are winners. Now the bars on that diagonal are the square ones: 1 × 1, 2 × 2, 3 × 3, 4 × 4. This suggests a simple strategy that should apply to bars of any size: squares are losers, rectangles are winners. Having noticed this apparent pattern, we can check its validity *without* working through the entire game tree by verifying properties (1) and (2). Here's the reasoning. Clearly any rectangle (winner) can always be converted to a square (loser) in one move. In contrast, whatever move you make starting with a square (loser), you cannot avoid leaving your opponent a rectangle (winner). Moreover, ■ is square, and we know it is a losing position. All this is consistent with principles (1) and (2), so working backwards we deduce (recursively) that *every* square is a loser and *every* rectangle a winner. We now see that Too's first move in Figure 65 was a mistake. And we see that yucky choccy is a dream game no matter what size the bar is.

In principle the same procedure applies to any finite game. The opening position is the 'root' of the game tree. At the other extreme are the tips of the outermost twigs, which terminate at positions where one or other player has won. Since we know the win/lose status of these terminal positions, we can work backwards along the branches of the game tree using principles (1) and (2), labelling positions 'win' or 'lose' as we proceed. The first time, we determine the status of all positions that are one move away from the end of the game. The next time, we determine the status of all positions that are two moves away from the end of the game, and so on. Since, by assumption, the game tree is finite, eventually we reach the root of the tree – the opening position. If this gets the label 'win' then Wun has a winning strategy; if not, Too has.

We can even say, again in principle, what the winning strategy is. If the opening position is 'win' then Wun should always move to a position labelled 'lose' – which Too will then face. Because this is a losing position, any move Too makes presents Wun with a 'win' position. So Wun can repeat the same strategy until the game ends. Similarly, if the opening position is labelled 'lose', then Too has a winning strategy – with the same description. So in finite, drawless games, working backwards through the game tree in principle decides the status of all positions, including the opening one. I say 'in principle' because the calculations become intractable if the game tree is large. And even simple games can have huge game trees, because the game tree involves all possible positions and all possible lines of play. This opens the door to nightmare games.

We now contrast yucky choccy with a game whose rules are almost the same, but where pruning the game tree rapidly becomes impossible – and where pruning *is* possible, it does not reveal any pattern that could lead to a simple strategic recipe. That game is chomp, invented many years ago by David Gale (University of California at Berkeley) and described in his marvellous book on recreational mathematics, *Tracking the Automatic Ant* (Further Reading). Gale describes chomp using a rectangular array of cookies or biscuits, but I'll stick to chocolate. (It is best played with an array of buttons or the like.) Chomp is just like yucky choccy, with the sole difference that a legitimate move consists of removing a rectangular chunk of chocolate, as in Figure 66. Specifically, a player chooses a component square and then removes all squares in that row and column, together with all squares to the right of and above these.

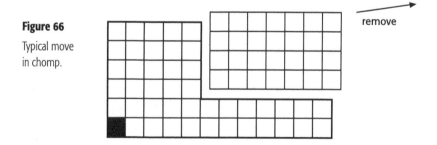

Figure 66

Typical move in chomp.

remove

There is a neat proof that for any size of bar (Figure 67a) other than 1 × 1, chomp is a win for Wun. Suppose, to the contrary, that Too has a winning strategy. Wun then proceeds by removing the upper right square (Figure 67b). This cannot leave Too facing a losing position, since we are assuming the opening position is a loser for Wun. So Too can play a winning move, something like Figure 67c, to leave Wun facing a loser. But then Wun could have played Figure 67d, leaving Too facing the same loser. This contradicts the assumption that Too has a winning strategy, so that assumption must be false. Therefore Wun has a winning strategy.

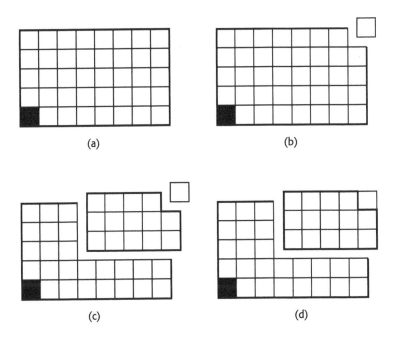

(a)   (b)

(c)   (d)

Figure 67

(a) Chomp bar ready for strategy stealing.

(b) If Wun does this ...

(c) ... and Too makes a supposed winning move ...

(d) ... then Wun could have played Too's move in the first place.

Proofs of this kind are called 'strategy stealing'. If Wun can make a 'dummy' move, pretend to be the second player, and win by following what ought to be a winning strategy for Too, then Too could not have had such a strategy to begin with – implying that Wun must have a winning strategy. The irony of this method of proof, when it works, is that it offers no clue to what Wun's winning strategy should be!

For chomp, detailed winning strategies are unknown, except in a few simple cases. In the $2 \times n$ (or $n \times 2$) case, Wun can always ensure that Too faces a position that is a rectangle minus a single corner square (Figure 68a). In the $n \times n$ case, Wun removes everything except an L-shaped edge (Figure 68b), and after that copies whatever move Too makes, but reflected in the diagonal. A few other small cases are known: for example in 3×5 chomp the sole winning move for Wun is Figure 68c. 'The' winning move need not be unique: in the 6×13 game there are two different winning moves.

Figure 68

Winning moves in

(a) $2 \times n$ chomp.

(b) $n \times n$ chomp.

(c) 3×5 chomp.

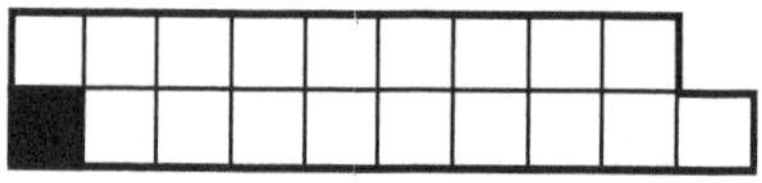

(a)

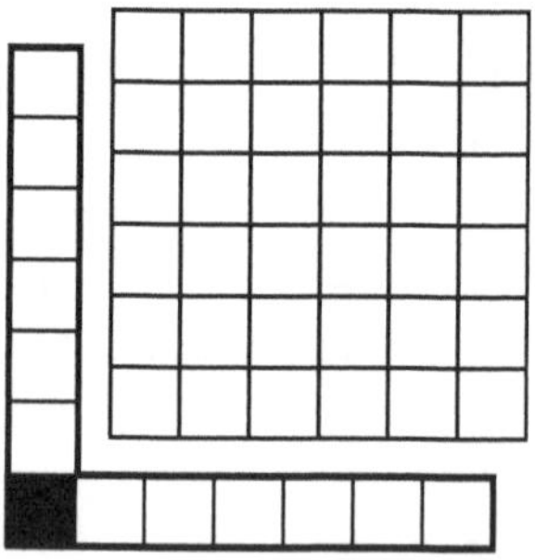

(b)

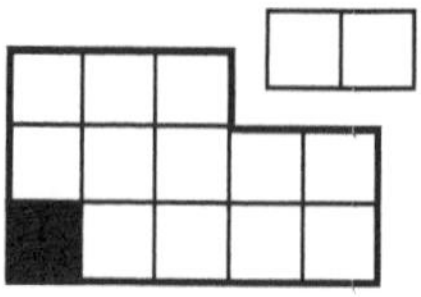

(c)

Other information about chomp positions can be found in *Winning Ways* by Berlekamp, Conway, and Guy (Further Reading). Chomp can also be played with an infinite chocolate bar – in which case, paradoxically, it remains a finite game because after finitely many moves only a finite portion of bar remains. But there is a change: Too can sometimes win. This happens, for example, with the 2 × ∞ bar. Figure 69 shows that whatever Wun does, Too can choose a reply that leads to Figure 68a, which we already know is a loser. Strictly speaking, I should be more careful here. By '∞' I really mean the set of positive integers in their usual order, which set theorists symbolize as ω ('omega') and refer to as 'the first infinite ordinal'. There are many other infinite ordinals, but their properties are too technical to describe here: see Gale's book for further details. Chomp can be played on doubly infinite arrays of ordinals, or in three or more dimensions: on the whole, little is known about winning strategies for these generalizations.

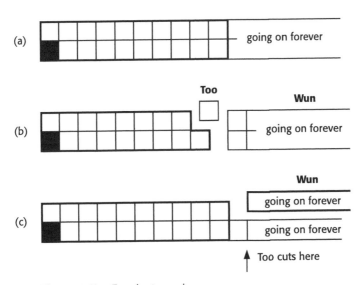

Figure 69 How Too wins 2 × ∞ chomp.

(a) Start.

(b) One type of possible play for Wun and its reply.

(c) The other type of possible play for Wun and its reply.

Shedding a Little Darkness

If two people stand in a hall of
mirrors, and one lights a match,
can the other always see it —
reflected in some series of mirrors,
possibly very, very dim, but visible
if it's bright enough? With curved
mirrors, the answer is 'no'. But what
if they are flat? In 1995, it was
proved that the answer is 'no' there,
too. Appropriately, the proof involves
a clever use of reflections.

Angela is standing in a hall of mirrors – a room whose walls are perfectly reflecting. Elsewhere in the room, her friend Bruno lights a match. Is it true that, no matter where they are standing, Angela can always see the match or one of its reflections, provided that she looks in the right direction? Equivalently, is the room 'illuminable from every point' in the sense that the light from the match fills the whole room – not even missing an isolated point – no matter where the match is placed?

This problem was first asked in print by Victor Klee in 1969, but its origins are thought to go back further, at least to Ernst Straus in the 1950s. It comes in several variants: the room may be merely a two-dimensional plan, or a genuinely three-dimensional shape: if the latter then its floor and ceiling – more generally, all of its interior surfaces – must also be mirrors. In either case we can ask the question for rooms with flat sides – polygons in two dimensions, polyhedrons in three – or with curved sides. In all versions of the problem the standard mathematical idealization replaces Angela's eye and Bruno's flame by points, *not* lying on the room's mirror-covered boundary, and both Angela and Bruno are assumed to be transparent. The law of reflection at any part of the boundary is the usual one: 'angle of incidence equals angle of reflection'. Those angles are defined only at points of the boundary that possess a unique tangent, so it is customary to assume that light that hits any boundary point without a unique tangent – such as a vertex of a polygon or polyhedron, a place where its boundary suddenly changes direction – is 'absorbed' and travels no further.

In two dimensions a negative answer, which I shall describe later in this chapter, was provided by L. Penrose and Roger Penrose in 1958. It required curved boundaries, and until very recently the problem for planar polygonal rooms remained open. Its solution was published by George Tokarsky in the December 1995 issue of *The American Mathematical Monthly* (see Further Reading). His elegant proof appropriately involves a 'reflection trick' and, like all the best mathematics, is astonishingly simple. The same kind of reflection trick is widely used in mathematics. Tokarsky generalizes his method to provide many other examples in two and three dimensions in which flat-sided rooms are not illuminable from every point.

The key idea is to start with an isosceles right triangle – a square divided in half along a diagonal, with one angle of 90° and two of 45°. Such a triangle can be 'unfolded' into a regular lattice pattern (Figure 70) by repeatedly reflecting it about its three sides. If you were to make a room whose floor plan was the same shape and whose walls were mirrors, and stood inside it, then the walls would have a kaleidoscope effect and you would 'see' this lattice pattern.

The lattice is used to prove a key fact: if a match is placed at one of the 45° angles of a room shaped like this triangle, with mirror walls, then no ray that emanates from it can return to the match, no matter how many times it bounces off the mirrors. To see why, first observe that any such ray, for example the one marked ABCD, can be unfolded in the same manner as the triangle. For example the segment BC inside the triangle unfolds to BC' on the other side of the wall, and continuing the process CD unfolds to C'D'. So ABCD here unfolds to give ABC'D'. The law of reflection implies that the unfolded ray ABC'D' is a straight line, and that fact is also crucial to all that follows.

We have coloured the three vertices of the triangle so that the vertex A at a 45° angle is black, the other 45° vertex is white, and the 90° vertex is grey. The illustrated path ABCD terminates at D because that is a corner of the triangle: equivalently, the 'unfolded' point D' lies in the lattice. We argue that if there were a path leading from A back to A,

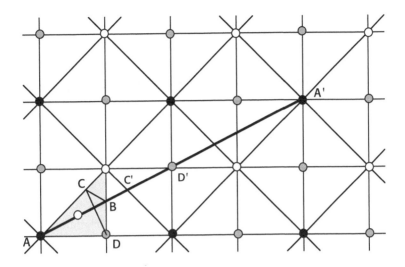

Figure 70 An isosceles right triangle (shaded) 'unfolds' by repeated reflection to form a lattice: dots show the correspondence between lattice points and vertices of the triangle. A light ray such as ABCD also unfolds, to give a straight line path ABC'D'. Any putative light ray from A back to A similarly unfolds to give a straight line such as AA'. However, this line must pass through a grey or white lattice point (here D') so the original ray must hit a vertex and be absorbed.

then the same thing would necessarily happen – implying that no such path can exist.

To prove this, imagine that we have some path from A to A and unfold it to get a straight line path from A to some point A' in the lattice. Because A' folds up to give A, it follows that A' is one of the black dots in the lattice. Now the black dots are spaced an *even* number of lattice units apart in both the horizontal and the vertical direction – their coordinates are even integers. This implies that somewhere along the line AA' there must be a lattice point that is either white or grey. This fact is obvious if either the horizontal or vertical spacing is twice an odd number, because then the midpoint of AA' has at least one coordinate an odd integer and is therefore a grey or white lattice point. This argument fails if both coordinates are multiples of four, but then the midpoint A″ of AA′ is a black lattice point and we can look at AA″ and try again. Either the midpoint of that is a grey or white lattice point, or A′ *also* has both coordinates multiples of 4. If the latter, we can replace A′ by the new midpoint A″, and so on. After finitely many such replacements, the first case of the argument must apply. For example, if the coordinates of A′ are 48 horizontally and 28 vertically then A′ has coordinates (24, 14), and A″ has coordinates (12, 7). The midpoint of AA″ is therefore a grey or white lattice point.

Having established that *any* unfolded path that joins A to a black lattice point must hit a grey or white lattice point, we now fold the path up again to conclude that the original path inside the triangle hits one of the other two corners (and is thus absorbed) before it gets back to A. That is what we wanted to establish.

We can construct polygonal rooms by fitting together horizontal, vertical, or diagonal segments of the lattice drawn in Figure 70. Suppose that a light ray bounces around inside such a room, starting at one black dot (Bruno), ending at a different black dot (Angela), and bouncing off the walls according to the equal-angle law of reflection. Then we can fold that path up to get a light path in the original equilateral right triangle that generated the lattice. However, we have just established that any such path will hit a grey or white vertex, and unfolding again we conclude that the original path must hit a grey or white lattice point. So any light ray that starts at Bruno, bounces off the mirror walls, and ends at Angela must encounter a grey or white dot along the way. Suppose we arrange for the following three conditions to hold:

- The two black dots (representing Angela and Bruno) are in the interior of the room.
- No grey or white dot lies in the interior of the room.
- Every grey or white dot on the boundary of the room lies at a vertex.

Then any ray that hits a grey or white dot must hit a vertex, and thus be absorbed – so there is no such light ray at all.

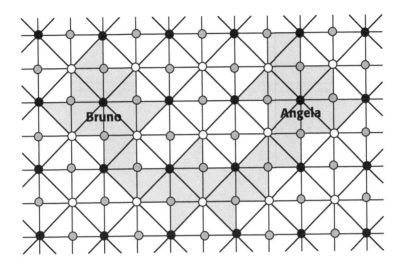

Figure 71 A hall of mirrors in which Angela cannot see Bruno's match.

An example of such a room is given in Figure 71. If you try to design such rooms you will find that it takes a certain amount of ingenuity to make sure that all three of the above rules are respected. It's easier than you might think to have grey or white lattice points on the boundary but not at a vertex, for example, so that you have to add extra triangles to introduce additional bends into the boundary; and unless you're careful these may create extra interior lattice points that you don't want because they violate the second rule ... But with a little care it's easy enough.

The room shown in Figure 71 is built up from 39 reflected copies of the isosceles right triangle plus the original. Tokarsky's article includes one with only 29 component triangles. Can you find it? Can anyone do better? What about minimizing the number of edges? Tokarsky also develops a similar theory for rooms obtained by 'unfolding' a square instead of an isosceles right triangle (Figure 72a), triangles of other shapes (Figure 72b), and three-dimensional rooms obtained using similar reflection principles.

Figure 72

Angela also cannot see Bruno's match in rooms created by reflecting

(a) squares and

(b) triangles whose angles are 9°, 72°, and 99°.

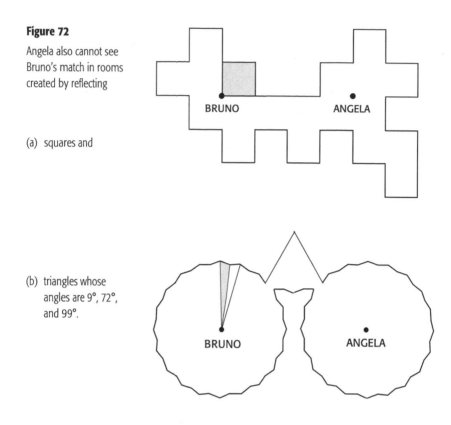

These examples show that in a polygonal room there may exist positions for a match that do not illuminate every point in the room. However, all we have proved is that at least one point is unilluminated. Is it possible for there to be a whole *region* of non-zero area that is not illuminated? This problem is distinctly trickier – for example in Figure 70 it is not clear, and we certainly have not proved, that rays starting from Bruno cannot pass as close as we wish to Angela. All we know is that they cannot hit her *exactly*. The answer seems not to be known for polygonal rooms, but if the room has curved sides then a clever argument due to the Penroses shows that unilluminated regions can

Figure 73

(a) A ray passing between foci of an ellipse is reflected back between them.

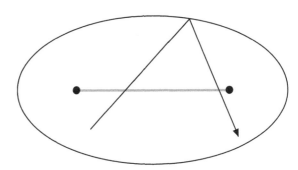

(b) Two half-ellipses with foci as shown, joined by wiggly curves whose precise shape is unimportant. No ray originating in region 'Bruno' can ever reach either 'Angela' region.

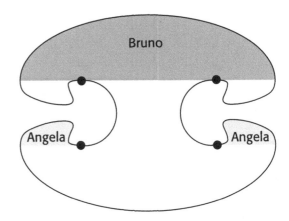

exist. Recall that the curve known as an ellipse has two special points, known as *foci* (Figure 73a). It can be proved that any light ray that passes between the two foci and bounces off the curve will again pass between the two foci before it next hits the curve. By 'pass between' the foci we mean 'cross the straight line that joins them'. Bearing this property in mind it is easy to check that the room in Figure 73b has un-illuminable regions. Specifically, rays originating in the shaded region labelled Bruno can never enter the lightly shaded one labelled Angela.

In a letter, Tokarsky sent me many more examples, and some generalizations. Among them are the following. There are polygonal rooms

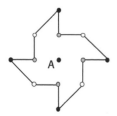

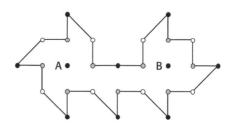

Figure 74

Room in which you cannot see
yourself reflected, if you are a
point at A.

Figure 75

Unilluminable room with 24 sides

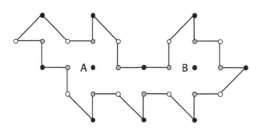

Figure 76

Unilluminable room with
27 sides, an odd number.

in which you cannot see yourself reflected (if you are a point), for
example Figure 74. There exists an unilluminable room with 24 sides,
fewer sides than the examples listed above (Figure 75). And although
all the examples of unilluminable rooms that I've shown you so far
have an even number of sides, there is one with 27 sides, an odd
number (Figure 76).

There are many similar problems, some solved, some not. J. Rauch
has shown that there exists a curved room, whose boundary has
a tangent at every point except one, that requires infinitely many
matches to illuminate every point. He has also shown that for any finite
number of matches there is a room, whose boundary has a tangent at
every point, that cannot be illuminated by that many matches. And
J. Pach has asked the very elegant question: if you light a match in
a forest of perfectly reflecting trees, must the light be visible from
outside? Here the trees can, for example, be modelled as circles, and
the problem is posed in the plane. The answer is not known.

Preposterous Piratical Predicaments

The pirates are fierce, but democratic. When they split the loot from their latest raid, the fiercest one proposes a way to divide it and everyone votes. If the vote is lost, the proposer walks the plank and the next fiercest pirate makes a new proposal, and so on. With ten pirates and 100 gold pieces, which proposal yields the fiercest pirate the biggest haul? And what if there are 500 pirates, but still only 100 gold pieces? The answers are amazing – and the fact that there are answers is even more so.

The logic of mathematics sometimes leads to apparently bizarre conclusions. The rule here is that *provided* the logic doesn't have holes in it, then the conclusions are sound. If they conflict with your intuition, then either your intuition is OK in a different context – but not the one under consideration – or you need to refine your intuition.

In September 1998 Stephen M. Omohundro sent me a puzzle that falls into exactly this category. The puzzle was invented by Steven Landsberg (University of Illinois, Urbana-Champaign), and Omohundro came up with a variant in which the logic becomes surprisingly convoluted and the conclusions are remarkable.

First, the original version of the puzzle.

Ten pirates have got their hands on a hoard of 100 gold pieces, and wish to divide the loot between them. They are democratic pirates, in their own way, and it is their custom to make such divisions in the following manner. The fiercest pirate makes a proposal about the division, and everybody votes – one vote each including the proposer. If 50% or more are in favour, the proposal passes and is implemented forthwith. Otherwise the proposer is thrown overboard and the procedure is repeated with the next fiercest pirate.

All the pirates enjoy throwing people overboard, but given the choice they prefer hard cash. They dislike being thrown overboard themselves. All pirates are rational, know that the other pirates are rational, know that they know that ... and so on. Unlike other puzzles with an apparently similar form (see Chapter 1), this one does *not* hinge upon the revelation of some piece of 'common knowledge'. Any common knowledge is already commonly known. Moreover, no two pirates are equally fierce, so there is a precise 'pecking order' – and it is known to them all. Finally: gold pieces are indivisible and arrangements to share pieces are not permitted (since no pirate trusts his fellows to stick to such an arrangement). It's every man for himself.

Which proposal will maximize the fiercest pirate's gain?

Omohundro's contribution is to ask the same question, but with 500 pirates instead of ten. (Still 100 gold pieces.) For convenience, number the pirates from 1 upwards in order of meekness, so that the *least* fierce is number 1, the next least fierce number 2, and so on. The fiercest pirate thus gets the biggest number, and proposals proceed in reverse numerical order from the top down.

Following Omohundro, I'm going to try to convince you that the answer to the ten-pirate version of the problem is this: Pirate 10 proposes to keep 96 gold pieces for himself, to give one gold piece to each of pirates 8, 6, 4, and 2, and none to the odd-numbered pirates. In contrast, the answer to the 500-pirate version involves the first 44 pirates being thrown overboard, after which pirate number 456, ... but I'm in danger of giving too much away, so you'll have to wait for the rest.

As we saw in Chapter 17, the secret to analysing all such games of strategy is to work backwards from the end. At the end, you know which decisions are good and which bad. Having established that, you can transfer that knowledge to the next-to-last decision, then the next-to-next-to-last, and so on. Starting from the front, in the order in which the decisions are actually taken, doesn't get you very far. The reason is that strategic decisions are all about 'What will the next person do if I do *this* ...?' so the decisions that follow yours are important. The ones that come before yours aren't, because you can't do anything about them anyway.

Bearing this in mind, the place to start is when (if!) the game gets down to just two pirates, P1 and P2. The fiercest pirate is P2, and (if the game ever gets this far) his optimal decision is obvious: propose

100 pieces for himself and none for P1. His own vote is 50% of the total, so it wins. Now add in pirate P3. Pirate P1 knows – and P3 knows that he knows – that if P3's proposal is voted down, P1 will get nothing. So P1 will therefore vote in favour of anything that P3 proposes, provided it yields him more than nothing. P3 therefore uses as little as possible of the gold to bribe P1, leading to the following proposal: 99 to P3, 0 to P2, and 1 to P1. Let's write it out like this:

P1 P2 P3
1   0   99

The thought-processes of P4 are similar. He needs 50% of the vote, so again he needs to bring exactly one other pirate on board. The minimum bribe he can use is one gold piece, and he can offer this to P2 since P2 will get nothing if P4's proposal fails and P3's is voted on. So the proposed allocation here becomes

P1 P2 P3 P4
0   1   0   99

The thought-processes of P5 are slightly more subtle. He needs to bribe *two* pirates. The minimum bribe he can use is two gold pieces, and the unique way he can succeed with this number is to propose the allocation

P1 P2 P3 P4 P5
1   0   1   0   98

The analysis proceeds in the same manner, with each proposal being uniquely prescribed by giving the proposer the maximum possible, subject to ensuring a favourable vote, until we get to the tenth pirate and find the allocation

P1 P2 P3 P4 P5 P6 P7 P8 P9 P10
0   1   0   1   0   1   0   1   0   96

This is the proposal I indicated earlier, and it solves the ten-pirate puzzle.

Now comes Omohundro's twist: what if there are (a lot) more pirates? Clearly the same pattern persists – for a while. In fact it persists up to the 200th pirate. P200 will offer nothing to the odd numbered pirates P1–P199, and one gold piece to each of the even-numbered pirates P2–P198, leaving one for himself.

However, we're trying to find the strategy for P500. At first sight, the argument breaks down at P200, because P201 has run out of bribes. However, he still has a vested interest in not being thrown overboard, so he can propose to take nothing himself:

P1	P2	P3	P4	...	P197	P198	P199	P200	P201
1	0	1	0	...	1	0	1	0	0

This opens up a new phase of the strategy, because pirate P202 *knows* that P201 has to accept nothing, or be thrown overboard. So he can count on P201's vote. However, P202 also is forced to accept nothing. He must use the entire 100 gold pieces to bribe 100 pirates – and these must be among those who would get zero according to P201's proposal. Since there are 101 such pirates, P202's proposal is no longer unique. Let's use a star to mark pirates who *might* get something from P202's proposal:

P1	P2	P3	P4	...	P197	P198	P199	P200	P201	P202
0	*	0	*	...	0	*	0	*	*	0

At this point a further consideration comes into force. Pirates have to think about how a pirate who has some *chance* of receiving a gold piece will react if he is definitely offered a gold piece. This depends on how much money, on average, he is willing to sacrifice for the fun of throwing somebody overboard. The puzzle doesn't specify this, so it is reasonable to assume that the pirates don't know it. This means that the bribe *might* succeed, so it is rational to offer a definite piece to a pirate who only has a chance of one later.

As it turns out, things are more satisfactory: there are enough definite 0s in each round for pirates from now on to offer the bribes only to *them*. Indeed, as Peter Norvig pointed out, there will always be 100 allocations of 0 among pirates P1–P200, so we may describe a solution in which bribes are offered only to (some of) them. You may wonder why I'm bothering about all this, since clearly pirate 203 doesn't have enough cash available to bribe enough pirates to vote for his proposal, whatever it is. This is true, and P203 goes overboard whatever he proposes – a result that we symbolize by a cross ×. We write ? to show that his choices are irrelevant, giving the situation:

P1	P2	P3	P4	...	P197	P198	P199	P200	P201	P202	P203
?	?	?	?	...	?	?	?	?	?	?	×

Even though P203 is destined to walk the plank, this doesn't mean that he plays no part in the proceedings. On the contrary, P204 now knows that P203's sole aim in life is to avoid having to propose a division of the spoils. So P204 can count on P203's vote, whatever P204 proposes. Now P204 just squeaks home – he can count on his own vote, P203's vote, and 100 others from bribes of one gold coin: 102 votes in all, the necessary 50%. So P204 will propose

P1	P2	P3	P4	...	P197	P198	P199	P200	P201	P202	P203	P204
*	0	*	0	...	*	0	*	0	*	*	0	0

What of P205? He is not so fortunate! He cannot count on the votes of P203 or P204: if they vote against him they have the fun of throwing him overboard and can still save themselves. So P205 gets a ×. So does P206 – he can be sure of P205's vote, but that's not enough. In the same way, P207 needs 104 votes – 3 plus his own plus 100 from bribes. He can get the votes of P205 and P206, but he needs one more and it's not available. So P207 gets a × as well. P208 is more fortunate. He also needs 104 votes, but P205, P206, and P207 will vote for him! Add in his own vote and 100 bribes and he's in business. He must offer bribes to those pirates that would get 0 according to P204's proposal:

P1	P2	P3	P4	...	P198	P199	P200	P201	P202	P203
0	*	0	*	...	*	0	*	0	0	*

P204	P205	P206	P207	P208
*	0	0	0	0

Now a new pattern has set in, and it continues indefinitely. Pirates who make proposals (always to give themselves nothing and to bribe 100 of the first 200 pirates) are separated from each other by ever-longer sequences of pirates who will be thrown overboard no matter what proposal they make – and whose vote is therefore ensured for any fiercer pirate's proposal. The pirates who avoid this fate are numbers P201, P202, P204, P208, P216, P232, P264, P328, P456, ... and so on – 200 plus a power of 2.

It remains to work out who are the lucky recipients of the bribes, just to make sure they will accept them. As I said, the solution is not unique, but one way to do this is for P201 to offer bribes to the odd-numbered pirates P1–P199, for P2 to offer bribes to the even-numbered pirates P2–P200, then P204 to the odd numbers, P208 to the evens, and so on, alternating from odd to even and back. At any rate, we conclude that with 500 pirates and optimal strategy, the first 44 pirates are thrown overboard, and then P456 offers one gold piece to each of P1, P3, P5, ... , P199.

Thanks to their democratic system, the pirates have arranged their affairs so that the very fierce ones mostly get thrown overboard, and at best can consider themselves lucky to escape with none of the loot. Only the 200 meekest pirates can expect anything, and only half of *them*. Truly, the meek shall inherit the worth.

Million-Dollar Minesweeper

It's not often you can win a million dollars by analysing a computer game, but by a curious conjunction of fate, there's a chance that you might. However, you'll only pick up the loot if all the experts are wrong and a problem that they think is extraordinarily hard turns out to be easy. So don't order the Ferrari yet.

The prize is one of seven now on offer from the newly founded Clay Mathematics Institute in Cambridge MA, set up by businessman Landon T. Clay to promote the growth and spread of mathematical knowledge, each bearing a million-buck price-tag. The computer game is Minesweeper, which is included in Microsoft's Windows™ operating system, and involves locating hidden mines on a grid by making guesses about where they are located and using clues provided by the computer. And the problem is one of the most notorious open questions in mathematics, which rejoices in the name 'P=NP?'.

The connection between the game and the prize problem was explained by Richard Kaye of the University of Birmingham, England (see Further Reading). And before anyone gets too excited, you won't win the prize by winning the game. To win the prize, you will have to find a really slick method to answer questions about Minesweeper when it's played on gigantic grids – and all the evidence suggests that there isn't a slick method. In fact, if you can *prove* that there isn't one, you can win the prize that way too.

Let's start with Minesweeper. The computer starts the game by showing you a blank grid of squares. Some squares conceal mines; the rest are safe. Your task is to work out where the mines are without detonating any of them. You do this by choosing a square. If there's a mine underneath it, the mine is detonated and the game ends – with a loss for you, of course. If there is no mine, however, the computer writes a number in that square, telling you how many mines there are in the eight immediately adjacent squares (horizontally, vertically, and diagonally).

If your first guess hits a mine, you're unlucky: you get no information except that you've lost. If it doesn't, though, then you get partial information about the location of nearby mines. You use this information to influence your next choice of square, and again either you detonate a mine and lose, or you gain information about the positions of nearby mines. If you wish, you can choose to mark a square as containing a mine: if you're wrong, you lose. Proceeding in this way, you can win the game by locating and marking all the mines.

For instance, after a few moves you might reach the position shown in Figure 77. Here a star shows a known mine (position already deduced), the numbers are the information you've got from the computer, and the letters mark squares whose status is as yet untested. With a little thought, you can deduce that the squares marked A must contain mines, because of the 2s just below them. The squares marked B must also contain mines, because of the 4s and 5s nearby. In the same way, C must contain a mine; and it then follows that D and E do not. The status of F can then be deduced, after a few moves, by uncovering D and seeing what number appears.

F	D	2	1	2	1
A	A	3	*	4	B
2	2	3	*	5	B
0	0	1	1	4	B
0	1	1	1	2	B
0	1	C	E	E	E

Figure 77
Typical Minesweeper position.

Now, the P=NP? problem. Recall that an algorithm is a procedure for solving some problem that can be run by a computer: every step is specified by some program. A central question in the mathematics of computation is: how efficiently can an algorithm solve a given problem? How does the running time – the number of computations needed to get the answer – depend on the initial data? For theoretical purposes the main distinction is between problems that are of type P – polynomial time – and those that are not. A problem is of type P if it can be solved using an algorithm whose running time grows no faster than (a constant multiple of) some fixed power of the number of symbols required to specify the initial data. Otherwise the problem is non-P. Intuitively, problems in P can be solved efficiently, whereas non-P problems cannot be solved algorithmically in any practical manner because any algorithm will take a ridiculously long time to get an answer. Problems of type P are easy, non-P problems are hard. Of course it's not quite as simple as that, but it's a good rule of thumb.

You can prove that a problem is of type P by exhibiting an algorithm that solves it in polynomial time. For example, sorting a list of numbers into numerical order is a type P problem, which is why commercial databases can sort data; and searching a string for some sequence of symbols is also a type P problem, which is why commercial word-processors can carry out search-and-replace operations. In 2003, to the surprise of many mathematicians, it was proved that testing whether a number is prime is a type P problem, requiring a computation that grows no faster than the 12th power of the number of digits (see Agrawal *et al.* and Bornemann, Further Reading).

In contrast, the Travelling Salesman Problem – find the shortest route whereby a salesman can visit every city on some itinerary – is widely believed to be non-P, but this has never been proved. Finding the prime factors of a given integer is also widely believed to be non-P, too, but this has never been proved either. The security of certain cryptosystems, some of which are used to send personal data such as credit card numbers over the Internet, depends upon this belief being correct.

Why is it so hard to prove that a problem is non-P? Because you can't do that by analysing any particular algorithm. You have to contemplate *all possible algorithms* and show that none of them can solve the problem in polynomial time. This is a mindboggling task.

The best that has been done to date is to prove that a broad class of candidate non-P problems are all on the same footing – if any one of them can be solved in polynomial time, then they all can. The problems involved here are said to have 'nondeterministic polynomial' running time: type NP.

NP is not the same as non-P. A problem is NP if you can check whether a proposed solution actually *is* a solution in polynomial time. This is – or at least, seems to be – a much less stringent condition than being able to find that solution in polynomial time. My favourite example here is a jigsaw puzzle. Solving the puzzle can be very hard, but if someone claims they've solved it, it usually takes no more than a quick glance to check whether they're right. To get a quantitative estimate of the running time, just look at each piece in turn and make sure that it fits the limited number of neighbours that adjoin it. The number of calculations required to do this is roughly proportional to the number of pieces, so the check runs in polynomial time. But you can't solve the puzzle that way. Neither can you try every potential solution in turn and check each as you go along, because the number of potential solutions grows much faster than any fixed power of the number of pieces.

It turns out that a lot of NP problems have 'equivalent' running times. Specifically, an NP problem is said to be NP-complete if the existence of a polynomial time solution for that problem implies that *all* NP problems have a polynomial time solution. Solve one in poly-nomial time, and you've solved them all in polynomial time. A vast range of problems are known to be NP-complete. The P=NP? problem asks whether types P and NP are (despite all appearances to the contrary) the same. The expected answer is 'no'. However, if *any* NP-complete problem turns out to be of type P – to have a polynomial time solution – than NP must equal P. We therefore expect all NP-complete problems to be non-P, but no one can yet prove this.

One of the simplest known NP-complete problems is SAT, the logi-cal satisfiability of a Boolean condition. Boolean circuits are built from logic gates with names like AND, OR, and NOT. The inputs to these circuits are either T (true) or F (false). Each gate accepts a number of inputs, and outputs the logical value of that combination. For instance an AND gate takes inputs p, q and outputs p AND q, which is T provided p and q are both T, and F otherwise. A NOT gate turns input T into output F and input F into output T. The SAT problem asks, for a given Boolean circuit, whether there exist choices of inputs that produce the output T. If this sounds easy, don't forget that a circuit may contain huge numbers of gates and have huge numbers of inputs.

★	2	★			
★	★	★			
				0	0
6				0	1

Figure 78

Impossible Minesweeper position.

The link to the computer game comes when we introduce the Minesweeper Consistency Problem. This is not to *find* the mines, but to determine whether a given state of what purports to be a Minesweeper game is or is not logically consistent. For example, if during the state of play you encountered Figure 78, you would know that the programmer had made a mistake: there is no allocation of mines consistent with the information shown. Kaye proves that Minesweeper is equivalent to SAT, in the following sense. The SAT problem for a given Boolean circuit can be 'encoded' as a Minesweeper Consistency Problem for some position in the game, using a code procedure that runs in polynomial time. Therefore, if you could solve the Minesweeper Consistency Problem in polynomial time, you would have solved the SAT problem for that circuit in polynomial time. In other words, Minesweeper is NP-complete. So, if some bright spark finds a polynomial-time solution to Minesweeper, or alternatively proves that no such solution exists, then the P=NP? problem is solved (one way or the other).

Kaye's proof involves a systematic procedure for converting Boolean circuits into Minesweeper positions. Here a grid square has state T if it contains a mine, and F if not. The first step involves not gates, but the wires that connect them. Figure 79 shows a Minesweeper wire. All squares marked x either contain a mine (T) or do not contain a mine (F), but we don't know which. All squares marked x' do the opposite of x. You should check that all the numbers shown are correct whether x is T or F. The effect of the wire is to 'propagate' the signal T or F along its length, ready to be input into a gate.

0	0	0	0	0	0	0	0	0	0	0	0	0	0
1	1	1	1	1	1	1	1	1	1	1	1	1	1
x	x'	x	x'	x	x'	x	x'	x'	x'	x	x'	x	x'
1	1	1	1	1	1	1	1	1	1	1	1	1	1
0	0	0	0	0	0	0	0	0	0	0	0	0	

Figure 79 A wire.

Figure 80 shows a NOT gate. The numbers marked on the block in the middle force an interchange of x and x' on the exit wire, compared to the input wire. The AND gate (Figure 81) is more complicated. It has two input wires U, V, and one output W. To establish that this is an AND gate, we assume that the output is T and show that both inputs have to be T as well. Since the output is T, every symbol t must indicate a mine and every t' a non-mine. Now the 3 above and below a_3 implies that a_2 and a_3 are mines, so a_1 is not a mine, so s is a mine. Similarly, r is a mine. Then the central 4 already has four mines as neighbours, which implies that u' and v' are non-mines, so u and v are mines – and this means that U and V have truth-value T. Conversely, if U and V

					1	1	1					
1	1	1	1	1	2	*	2	1	1	1	1	1
x'	x	1	x'	x	3	x'	3	x	x'	1	x	x'
1	1	1	1	1	2	*	2	1	1	1	1	1
					1	1	1					

Figure 80 A NOT gate.

U																							
1	1	1			1	2	2	1		1	1	1		1	1	1							
1	u'	1			2	*	*	3	2	3	*	2	1	2	*	3	2	1					
1	u	1	1	2	4	*	s	a_1	a_2	a_3	t'	3	t	t'	3	*	*	2					
1	2	2	1	1	*	*	4	*	3	2	3	*	2	1	1	2	t	*	2				
2	*	u'	2	2	4	s'	3	1	1	0	1	1	1	0	0	1	2	2	1				
2	*	*	3	u	u'	s	2	1	1	1	1	1	1	1	1	1	t'	1	1	1	1	1	
2	4	5	*	4	*	4	t	t'	1	t	t'	1	t	t'	1	t	2	t	1	t'	t	1	
2	*	*	3	v	v'	r	2	1	1	1	1	1	1	1	1	1	t'	1	1	1	1	1	
2	*	v'	2	2	4	r'	3	1	1	0	1	1	1	0	0	1	2	2	1				W
1	2	2	1	1	*	*	4	*	3	2	3	*	2	1	1	2	t	*	2				
1	v	1	1	2	4	*	r	b_1	b_2	b_3	t'	3	t	t'	3	*	*	2					
1	v'	1			2	*	*	3	2	3	*	2	1	2	*	3	2	1					
V 1	1	1			1	2	2	1		1	1	1		1	1	1							

Figure 81 An AND gate.

have value T then so does *W*. In short, we have an AND gate as claimed.

There's more to Minesweeper electronics than this – for example, we need to be able to bend wires, split them, join them, or make them cross without connecting. Kaye solves all these problems, and other more subtle ones, in his article. The upshot is that solving the Minesweeper Consistency Problem is algorithmically equivalent to the SAT problem, and is thus NP-complete. To virtually every mathematician and computer scientist, this means that the Minesweeper Consistency

Problem must be inherently hard. It is astonishing that such a simple game should have such intractable consequences, but mathematical games are like that.

If you're interested in those million-dollar prizes, a word of warning. The Clay Institute imposes strict rules before it will accept a solution as being valid. In particular, it must be published by a major refereed journal, and it must have been 'generally accepted' by the mathematical community within two years of publication. But even if you're not going to tackle anything as daunting as that, you can have a lot of fun playing Minesweeper, secure in the knowledge that it encompasses one of the great unsolved problems of our age.

Further Reading

Chapter 1: I Know That You Know That ...

David Gale, More Paradoxes: Knowledge Games, *Mathematical Intelligencer* vol. 16 no. 4 (1994) 38–44.

J.M. Lasry, J.M. Morel, and S. Solimin, On knowledge games, *Revista Matematica de la Universidad Complutense de Madrid* vol. 2 (1989).

Chapter 2: Domino Theories

Martin Gardner, *Mathematical Puzzles and Diversions from Scientific American*, Bell, London 1961.

Russ Honsberger, *Mathematical Gems I*, Mathematical Association of America, Washington DC 1973.

Maurice Kraitchik, *Mathematical Recreations*, Allen and Unwin, London 1943.

Chapter 3: Turning the Tables

Elwyn R. Berlekamp, John H. Conway, and Richard K. Guy, *Winning Ways* vol. 2, Academic Press, New York 1982.

Chapter 4: The Anthropomurphic Principle

Robert Matthews, Tumbling toast, Murphy's Law, and the fundamental constants, *European Journal of Physics* vol. 18 (1995) 172–176.

Robert Matthews, The science of Murphy's Law, *Scientific American*, (April 1997), 72–75.

Chapter 5: Counting the Cattle of the Sun

Albert H. Beiler, *Recreations in the Theory of Numbers*, Dover, New York 1964.

David Fowler, *The Mathematics of Plato's Academy*, Oxford University Press, 1987.

Harry L. Nelson, A solution to Archimedes' Cattle Problem, *Journal of Recreational Mathematics* vol. 13 (1981) 162–176.

C. Stanley Ogilvy, *Tomorrow's Math* (2nd ed.), Oxford University Press, New York 1972.

Ilan Vardi, Archimedes' Cattle Problem, *American Mathematical Monthly* vol. 105 no. 4 (April 1998) 305–319.

Chapter 6: The Great Drain Robbery

Hallard T. Croft, Kenneth J. Falconer, and Richard K. Guy, *Unsolved Problems in Geometry*, Springer-Verlag, New York 1991.

H. Joris, Le chasseur perdu dans la forêt, *Elementa Mathematicae* vol. 35 (1980) 1–14.

Chapter 7: Two-Way Jigsaw Puzzles

Greg N. Frederickson, *Dissections: Plane and Fancy*, Cambridge University Press, Cambridge 1997.

Harry Lindgren, *Geometric Dissections*, Van Nostrand, Princeton 1964.

Ian Stewart, *From Here to Infinity* , Oxford University Press, Oxford 1996.

Stan Wagon, *The Banach–Tarski Paradox*, Cambridge University Press, Cambridge 1985.

Chapter 8: Tales of the Neglected Number

Mario Livio, *The Golden Ratio*, Broadway, New York 2002.

Benjamin M. M. de Weger, Padua and Pisa are exponentially far apart, *Publicacions Matemàtiques* vol. 41 (1997) 631–651.

Chapter 11: Is Monopoly Fair?
Nachum Dershowitz and Edward M. Reingold, *Calendric Calculations: the Millennium Edition*, Cambridge University Press, Cambridge 2001.

Chapter 12: Dividing up the Spoils
Steven J. Brams and Alan D. Taylor, An envy-free cake division protocol, *The American Mathematical Monthly* vol. 102 (Jan. 1995) 9–18.
David Gale, Mathematical entertainments, *The Mathematical Intelligencer* vol. 15 no. 1 (1993) 50–52.

Chapter 13: Squaring the Square
R. L. Brooks, C. A. B. Smith, A. H. Stone, and W. T. Tutte, The dissection of rectangles into squares, *Duke Mathematical Journal* vol. 7 (1940) 312–340.
Hallard T. Croft, Kenneth J. Falconer, and Richard K. Guy, *Unsolved Problems in Geometry*, Springer, New York 1991, p. 81.
David Gale, Mathematical entertainments, *The Mathematical Intelligencer* vol. 15 no. 1 (1993) 48–50.
Martin Gardner, *More Mathematical Puzzles and Diversions*, Bell, London 1963.

Chapter 14: The Bellows Conjecture
W. W. Rouse Ball, *Mathematical Recreations and Essays*, Macmillan, London 1939.

Chapter 15: Purposefully Piling Pyramids
Stuart Kirkland Wier, Insight from geometry and physics into the construction of Egyptian Old Kingdom pyramids, *Cambridge Archaeological Journal* vol. 6 (1996) 150–163.

Chapter 16: Be a Dots-and-Boxes Grandmaster

Elwyn Berlekamp, *The Dots and Boxes Game*, A. K. Peters, Natick MA 2000.

Chapter 17: Choosily Chomping Chocolate

Elwyn R. Berlekamp, John H. Conway, and Richard K. Guy, *Winning Ways*, Academic Press, New York 1982, p. 598.

David Gale, *Tracking the Automatic Ant*, Springer, New York, 1998.

Chapter 18: Shedding a Little Darkness

George Tokarsky, Polygonal rooms not illuminable from every point, *American Mathematical Monthly* vol. 102 no. 10 (1995) 867–879.

Hallard T. Croft, Kenneth J. Falconer, and Richard K. Guy, *Unsolved Problems in Geometry*, Springer, New York 1991.

Victor Klee and Stan Wagon, *Old and New Unsolved Problems in Plane Geometry and Number Theory*, Mathematical Association of America, Washington DC 1991.

Chapter 20: Million-Dollar Minesweeper

Manindra Agrawal, Neeraj Kayal, and Nittin Saxena, PRIMES is in P, IIT Kanpur preprint August 8 2002;
http://www.cse.iitk.ac.in/news/primality.html

Folkmar Bornemann, PRIMES is in P: a breakthrough for "Everyman", *Notices of the American Mathematical Society* vol. 50 no. 5 (2003) 545–552.

Richard Kaye, Minesweeper is NP-complete, *Mathematical Intelligencer* vol. 22 no. 2 (2000) 9–15.

Index

Abbot, Stephen, 111
Agrawal, Manindra, 221
Airy, George Biddle, 80
allocation, 128–30, 137, 139
Amthor, A., 52
Andreas, J. M., 159
anthropomurphic principle, 44
Archimedes, 49, 51, 52, 54, 55, 161
  cattle problem, 49
Austin, Keith, 189

Banach, Stefan, 76, 78, 134, 135
Banach–Tarski paradox, 76
Battle of Brighton, 50
Battle of Hastings, 49
Beiler, Albert H., 50, 52
Bell, A. H. , 53
bellows conjecture, 160, 162, 164
Berlekamp, Elwyn, 179, 184, 185, 197
biped, 42
Bolyai, Wolfgang, 78, 89
Bolyai–Gerwein Theorem, 78, 89
Boolean, 222
Boolean circuit, 223, 224
Bornemann, Folkmar, 221
Boulanger, Philippe, viii
Bouwkamp, J. W., 149, 150

Bradley, Harry, 82
Brams, Steven, 137–40
Bricard, Raoul, 159, 160
Brooks, R. L., 147–9
Busschop, Paul, 81
Butler, William, 110
butterfly effect, 126

calendar, 119–21, 126
  Chinese, 121, 124, 125
  Gregorian, 121, 122
  Hebrew, 121
  Hindu, 124
  lunar, 120
  lunisolar, 124
  solar, 120
  Roman, 119
Carson, David, 44
cattle of the sun, 49, 51–4
Cauchy, Augustin-Louis, 159
Century puzzle, 34
Century and a Half puzzle, 34
chaos, 126
Chapman, S. J., 151
chomp, 189, 194–7
chord, 64–6, 69
Clay, Landon T., 219
Clay Mathematics Institute, 219, 226

Collison, David, 79
common knowledge, 5–7, 211
Connelly, Robert, 157, 159, 161, 163
Conway, John Horton, 6, 136, 137, 139, 197
critical overhang, 40–2
cube, 78
cubic equation, 91
cylinder, 150, 152

Dad's puzzler, 29, 34
Darrow, Charles, 116
Dehn, Max, 78, 145
Dershowitz, Nachum, 119, 121, 122
Dewdney, Alexander Keewatin, viii
dissection, 75, 77–82
division of cake, 126, 132
domino, 14–18, 20–1, 23, 182, 183
Donkey puzzle, 34
dots and boxes, 179, 185
double cross, 184
double-dealing, 182, 183
dream game, 189, 193
Dubins, L.E., 135
Dudeney, Henry Ernest, 49, 51, 75, 79, 83
Duivestijn, A. W. J., 149, 150

eigenvalue, 105, 109
eigenvector, 105, 109
Einstein, Albert, 44
electrical circuit, 147
electric charge, 103
ellipse, 207

energy, 168, 170
envy-free, 128, 130, 131, 133–9
epoch, 121, 123
Eratosthenes, 51
Euclid, 157
Euler, Leonhard, 50, 53

Fermat, Pierre de, 50
Fibonacci, 89, 91
    number, 87–91
Fish, E., 53
fixed date, 121
flexible, 158, 160
floor function, 122
focus, 207
Fowler, David, 54
Frederickson, Greg. N., 75, 83
Friddell, Thomas H., 110, 114

Gale, David, 6, 9, 150, 194
Galvin, Fred, 139
game tree, 190, 191, 193, 194
Gardner, Martin, vii–ix, 136, 147
German, R. A., 53
Gerrell, Spike, x
Gerwein, P., 78, 89
golden number, 87, 88, 91
Gomory, Ralph, 20,22, 23
great pyramid, 167
Gregory XIII, Pope, 121, 122
gridlock, 180–2
Guy, Richard K., 197

half-hearted handout, 185
Herodotus, 167

Heron of Alexandria, 161
Heron's formula, 161, 162
Hilbert, David, 78
Hofstadter, Douglas, vii
Holmes, Sherlock, 59–71
Homer, 51

illuminable, 201
induction, 6
irrational number, 120

jigsaw puzzle, 222
Joris, H., 68
Julian day, 121
Julius Caesar, 119

Kaye, Richard, 219, 223–5
Khayyam, Omar, 123
Khufu, 167–70, 174
Kirchhoff's laws, 147
Klee, Victor, 201
Klein bottle, 151–3
Knaster, B., 134, 135

Landsberg, Steven, 211
Laskar, Jacques, 126
lattice, 202–5
leap
   day, 121
   month , 126
   year, 122, 123
Leonardo of Pisa, 90
Lessing, Gotthold Ephraim, 49, 52, 54
Liapunov convexity theorem, 135

Lindgren, Harry, 80
Livio, Mario, 87
logarithmic spiral, 88
long chain, 183, 184
loony endgame, 185
loony move, 185
Loyd, Sam, 75, 76, 79, 83
Lucas, Édouard, 92
lunar month, 120

map, 29, 32–5
Markov, Andrey Andreyevich, 97, 103–5, 112
Markov chain, 97, 98, 109–12
Matthews, Robert, 39, 40, 42–4
maze, 30, 61
millennium bug, 119
minesweeper, 219, 223–6
mirror, 201
Möbius band, 150–2
modulo, 122
modulus, 122
Monopoly, 97, 100, 104, 109–11, 114–16
Moore, Alan, 174
Moore, Thomas, 39
Morón, Z., 145, 147, 152
Moskowitz, Bruce, 115
moving knife algorithm, 135
murphic resonance, 42
Murphy's Law, 39, 40, 42–5, 110

nautilus, 88
Nelson, Harry L., 53
Newton's laws of motion, 40, 41

nightmare game, 189, 194
nim-addition, 184
nimstring, 184, 185
non-P, 221
non-primality test, 92
NP, 222
NP-complete, 222, 223, 225
Nygren, A., 55

octahedron, 163
Olivastro, Dominic, 137
Omohundro, Stephen M., 211, 212, 214

P=NP?, 219, 220, 222, 224
Pach, J. , 208
Paddon, Earl A., 114
Padovan, Richard, 87, 89, 91
Padovan
  sequence, 89, 90
  number, 91, 92
parity, 181, 182
Paterson, Michael, 6
Payn, James, 39
Pell, J., 89
Pell equation, 51, 53, 89
  negative, 54
Penrose, L., 202, 206
Penrose, Roger, 202, 206
Perrin, R., 92
Perrin number, 90, 92, 93
pigeonhole principle, 126
pirate, 211–16
plastic number, 87, 88, 91
polyhedron, 157–61, 201

polymurph, 43
polynomial, 161
Press, W.H., 42
probability, 97, 100–2, 104, 110–13
projective plane, 151–3
proportional, 128, 130, 131, 133, 135
protocol, 128, 130, 131, 133–40
pruning, 191, 192
pyramid, 167–75
Pythagoras's Theorem, 80

quantum physics, 45
quintic equation, 91

rata die, 121
Rauch, J., 208
reflection trick, 202
Reingold, Edward M., 119, 121, 122
Richard, G. H., 53
Richey, Matt, 111
rigid, 157, 163
Rorres, Chris, 54, 55
Rouse Ball, W. W., 159

Sabitov, Idzhad, 157, 161, 163
SAT, 222, 223
sedan chair puzzle, 75, 76
Selfridge, John, 136, 137, 139
Séquin, Carlo, 44
Shallit, Jeffrey, 92
Simon, Jonathan, 116
simple, 147
sliding block puzzle, 34
smart Alec puzzle, 83

Smith, C. A. B., 147
Smith diagram, 147, 149
snakes and ladders, 104
solar day, 120
solitaire, 81
Sosigenes, 119
Spanier, Edwin, 135
Sprague, R., 145, 148
square, 52, 53
squared rectangle, 145, 147
squaring the circle, 144
squaring the square, 144, 145, 147, 149
Stalker, R. M., 159
Steadman, John, 44
Steffen, Klaus, 160
Steinhaus, Hugo, 132–4
step principle, 78
St. George, Alan, 87
Stone, A. H., 147
strap, 65, 66, 68
strategy, 131, 193, 212
strategy stealing, 195, 196
Straus, Ernst, 201
strip principle, 80, 81
sub-puzzle, 30–2, 33
Sullivan, Dennis, 160
symmetry, 103, 109, 110, 149
synodic month, 120

tangent, 64–6, 69, 201
Tarski, Alfred, 76, 78
Taylor, Alan, 138–40
tesselation principle, 79

tetrahedron, 78, 162
Theobald, Gavin, 77
Tilson, Philip, 77
toast, 39–42, 44, 110
Tokarsky, George, 202, 205, 207
torus, 150, 152
transition matrix, 103, 105, 109, 110
trench, 64, 69
triangular number, 52, 53
tropical year, 120
Tutte, W. T., 147
type P, 221

Vardi, Ilan, 49, 52, 53

Wadlow, Robert, 44
Wagon, Stan, 77
Wallace, William, 78
Walz, Anke, 157, 163
Washington, George, 192
Watson, Dr, 59–71
Weger, Benjamin de, 91
Weiblen, David, 114, 116
Wier, Stuart Kirkland, 167, 168, 170
Willcocks, T. H., 149
Williams, H.C., 53
winning strategy, 190, 191, 193, 194
Wisdom, Jack, 126
Wurm, J. F., 53

yucky choccy, 189, 191, 193, 194

Zarnke, C. R., 53
Zwicker, William, 139